The Origins of the Art and Practice of Professional Writing

SUNY series, Studies in Technical Communication
―――――――
Miles A. Kimball, Charles H. Sides, Derek G. Ross, and Hilary A. Sarat-St. Peter, editors

The Origins of the Art and Practice of Professional Writing

The Written Word as a Tool for Social Justice Then and Now

KATHRYN ROSSER RAIGN

Cover art by Jennifer Welden Evans

Published by State University of New York Press, Albany

© 2024 State University of New York

All rights reserved

Printed in the United States of America

No part of this book may be used or reproduced in any manner whatsoever without written permission. No part of this book may be stored in a retrieval system or transmitted in any form or by any means including electronic, electrostatic, magnetic tape, mechanical, photocopying, recording, or otherwise without the prior permission in writing of the publisher.

For information, contact State University of New York Press, Albany, NY
www.sunypress.edu

Library of Congress Cataloging-in-Publication Data

Name: Raign, Kathryn Rosser, author.
Title: The origins of the art and practice of professional writing : the written word as a tool for social justice then and now / Kathryn Rosser Raign.
Description: Albany : State University of New York Press, [2024] | Series: SUNY series, Studies in Technical Communication | Includes bibliographical references and index.
Identifiers: ISBN 9781438497280 (hardcover : alk. paper) | ISBN 9781438497303 (ebook) | ISBN 9781438497297 (pbk. : alk. paper)
Further information is available at the Library of Congress.

10 9 8 7 6 5 4 3 2 1

Because the messenger's mouth was too heavy, and he could not repeat it, the lord of Kulab patted some clay and put the words on it as on a tablet. Before that day, there had been no putting words on clay . . .

—Halton and Svard, *Women's Writing of Ancient Mesopotamia*

I want to dedicate this book to my husband Jerry, who first introduced me to the art of Mesopotamia. He listened patiently as I talked through my ideas, and he always believed in me. Thanks to my wonderful children, Erin and Max, because you never doubted that what I was doing was important. Linda and Michael, you were there the day I saw my first cylinder seal. And thanks to my large and extended family of colleagues, and friends. What would I do without you?

And I want to thank the JP Morgan Library. It was your collection of artifacts from ancient Mesopotamia that inspired this book. The three days I spent in your research library were a revelation. Without your generosity in sharing your collection, I couldn't have written this book.

Contents

List of Illustrations   xi

Acknowledgments   xiii

Preface   xv

Prologue   xvii

Introduction: Expanding the History and Purpose of
Technical Communication   1

Chapter 1   Cylinder Seals: Written Communication's First
Technological Breakthrough   13

Chapter 2   Ancient Technical Manuals and Letters: The Origins
of Instructional Writing   37

Chapter 3   Finding Our Missing Pieces: Women Technical
Writers in Ancient Mesopotamia   61

Chapter 4   Decentering the History of the Writing Center:
A Case for the Mesopotamian *Edubba* as
an Early Writing Center   89

Chapter 5   Mythos, Nomos, Logos: Evidence of Sophistic
Reasoning before the Sophists   113

Chapter 6 Myth, Magic, and Medicine: Medical Writing in
            Ancient Mesopotamia   133

Chapter 7 Writing as Social Justice   147

Notes   161

Bibliography   163

Index   171

List of Illustrations

Figures

I.1	Eme-bal, language turner.	7
I.2	Disk tokens from Uruk.	8
1.1	Cylinder seal of Kalki the Scribe.	14
1.2	Great Seal of Henry the Eighth.	16
1.3	Contest scene.	16
1.4	President Barack Obama's signature.	18
1.5	Large clay bulla.	19
1.6	Presentation scene.	23
1.7	Seal owned by slave girl.	24
1.8	Bull pulling plow.	25
1.9	Woman with wheat.	25
1.10	Seal of Queen Puabi.	27
1.11	Seal of Ibbi Sin.	30
1.12	Goddess leading a worshiper to a seated deity.	31
1.13	Hero fighting lion.	32
1.14	Scribe.	34
1.15	Winged hero.	35

4.1	Lentil-shaped tablet.	110
4.2	Small one-column Cuneiform tablet.	110
4.3	Cuneiform tablet.	111
4.4	Cuneiform cylinder with inscription of Nebuchadnezzar II.	111
6.1	Medical text.	139

Tables

I.1	Timeline of Major Mesopotamian Languages and Periods	9
2.1	Linguistic Structure of Instructions	40
2.2	Number of Tasks Per Line	46
4.1	Lexical List of Professions—Used for Scribal Education	96
4.2	Analysis	98
4.3	Types of School Tablets	110
5.1	Hoe's Use of Emotional Language	126
6.1	Use of Lines and Columns in Medical Text	140

Acknowledgments

I thank the journals that previously printed some of the material from this book: "The Art of Ancient Mesopotamian Technical Manuals and Letter: The Origins of Instructional Writing," Kathryn Rosser Raign, *Technical Communication Quarterly*, copyright © 2022 Association of Teachers of Technical Writing, reprinted by permission of Taylor & Francis Ltd, http://www.tandfonline.com on behalf of © 2022 Association of Teachers of Technical Writing; Kathryn R. Raign, "Finding Our Missing Pieces—Women Technical Writers in Ancient Mesopotamia," *Journal of Technical Writing and Communication*; and Kathryn Raign, "De-Centering the History of the Writing Center: A Case for the Mesopotamian Edubba as an Early Writing Center," *College, Composition and Communication Journal*.

Preface

I would like to begin with a story—my story. When I was hired for my first tenure-track position, I was hired by the English department to work in two fields: rhetoric and composition, and technical and professional communication (TPC). To avoid confusion, my contract stated that I would devote 50 percent of my workload to each—no more no less. For every rhetoric course I taught, I had to teach a technical communication course. For every article I published in a rhetoric and composition journal, I had to publish one in a TPC journal. I was neither fish nor fowl.

Several years after I had been in my job, my TPC colleagues decided it was time for the TPC program to move out on its own, so it could grow unfettered. We became a separate department, and I was forced to choose. The two departments, like divorced parents, were also forced to make choices. Which department kept the writing center? Which department could offer science writing? Would the Technical Communication Department be allowed to teach any rhetorical principles in their courses?

Because I had spent my career with my feet in both worlds, I was torn. Although I had been expected to live my academic life as though I were two people, that hadn't happened. Instead, working in both fields simultaneously had shown me that the two disciplines I loved were not separate: they had a symbiotic relationship—or at least they should have.

My opinion was not shared on either side. My friends in rhetoric and composition shuddered when the word "professional" was applied to writing. Their job was much loftier than merely teaching students communication skills they could monetize after graduation. My colleagues in my new department of technical communication rolled their eyes at the thought of teaching "theory," or those soft topics like social justice. We were social *scientists*, not humanists!

In my experience this battle is still waging in many universities. Both fields are already lower on the academic food chain than they should be, and their competition just works to keep them there. Rather than criticizing each other for our differences, why not embrace the commonalities to the benefit of both us and our students?

Humans are inherently persuasive. Since humans first learned to speak, they began to persuade. No one disputes this. What is disputed is who can claim it first. Was it the Greeks? The Sophists? Did it happen even earlier?

In this book, I explore a body of texts older than any yet identified by either the field of TPC, or rhetoric and composition. These ancient Mesopotamian documents, which were carved on clay tablets, demonstrate that as writing developed humans consciously applied principles of persuasion to the documents that they produced, and these documents include everything from receipts, to laws, to letters, to instructions, and to literature. Each genre developed because it served a purpose for the emerging society. I originally began my study with the intent of showing that professional communication developed first. Instead, I learned that as writing first developed, the overlapping threads that included rhetorical practices, the application of design, and the awareness of style for writing documents of all types, resulted in a fabric that includes both disciplines, not distinctly, but symbiotically. It is this relationship I explore in this book with the intention of showing that our disciplines are not and should not be separate. Instead, we should share our strengths, embrace our commonalities, and honor our differences.

Prologue

This book had an unexpected beginning. I was in New York City for a long weekend with my husband Jerry, my brother Michael, and his wife Linda. Our plans were fluid. We had come together for a long weekend of food, talk, and exploration. Michael and Linda, who live in Boston, spend a great deal of time in New York, while Jer and I, who live in Texas, were still discovering the city's many secrets.

According to my family, I never learned to read: I was born reading. With this in mind, Jerry said we had to go to the JP Morgan library, so I could indulge my bibliophilia in an orgy of some of the world's most famous texts. I expected to see the Gutenberg Bible, and I basked in the special collection of works by Lewis Carroll. But I made my discovery in a small back room of the library where few people ventured.

The room was small, dark, and lined with softly lit glass cases filled with small pieces of stone roughly the size and shape of a tube of pasta. At that point in our visit, I was tired, saturated with words, and my feet hurt. I was only in the room because my museum-going training insisted I leave no room unexplored. My dad would have been proud.

I'll admit I did not give the artifacts in the room the care or attention they deserved. I had never heard of cylinder seals, knew very little about ancient Mesopotamia, and didn't give them much thought. But I looked at them, wondered about their meaning, and read the printed explanations next to each case. Interestingly, people rolled these little tubes over wet clay to "seal" a transaction. I was long familiar with the expression, "seal the deal," but I'd had no idea of its origins. The room wasn't a waste after all. I could check another box on my list of interesting word-related facts.

That should have been it. But for the rest of the weekend, those little tubes niggled away at the back of my brain. I study technical communication, and I have for many years. I've delved into the history of my profession, but I'd never ventured past the ancient Greeks. By the time I got home, the niggle had become more insistent. Did the ancient Mesopotamians really create the first technology that allowed someone to mass produce their "signature"?

That question led me down a rabbit hole into the ancient past, and this book is the result.

Introduction

Expanding the History and Purpose of Technical Communication

> Because the messenger's mouth was too heavy, and he could not repeat it, the lord of Kulab patted some clay and put the words on it as on a tablet. Before that day, there had been no putting words on clay . . .
>
> —Halton and Svard, *Women's Writing of Ancient Mesopotamia*

You're sitting in a classroom listening to your instructor talk about the history of technical and professional writing. They seem to think this will make you, a sophomore taking a required course, more invested in what you'll be learning. Wow. People were actually writing instructions and memos before we had computers and the internet, so tech writing must be super old. Like, it started in the 1950s.

Where does the history of technical and professional communication (TPC) begin? With the Egyptians, or the ancient Greeks? During the Renaissance? With the Industrial Revolution? Logically, we should begin our count from the invention of the first form of writing, Cuneiform, which began in Mesopotamia in the final centuries before 3000 BCE and ended with the conquest of Babylon by Cyrus the Great of Persia in 539 BCE—2,500 years. "From 500 BCE to the present is the same distance in time" (Kriwaczek 2012, 244), yet Cuneiform and two dominant languages—Sumerian and Akkadian—were maintained throughout, though Sumerian ceased to be spoken. And the other dominant language, Assyrian, splintered into different dialects. The last recorded Assyrian emperor, Ashurbanipal (685–627 BCE), bragged about his ability to read

2 | The Origins of the Art of Professional Writing

"the cunning tablets of Sumer, and the dark Akkadian language which is difficult rightly to use; I took my pleasure in reading stone inscribed before the flood" (Kriwaczek 2012, 11).

Before the Greeks, King Solomon, Moses, Abraham, even Noah and his flood, people were settling in villages and imagining cities (Kriwaczek 2012) and using Cuneiform to do so. Cuneiform remained in use for 2,500 years, yet neither discipline has investigated the influence it and the languages it recorded have had on the past, present, and future of TPC.

The piece of writing from this period that most people have likely heard of is *The Epic of Gilgamesh*. Certainly, this is a story worth reading because it reveals so much about the beliefs, practices, and opinions of the people it represents, yet it is a poor choice to represent the writing of the age. Gilgamesh is a work of literature, and the Mesopotamian literary canon contains other great works of literature. However, most of the documents that have been recovered and translated were not works of literature. Instead, they were the writing of average people conducting mundane business:

> Tell the Lady Zinu: Iddin-Sin sends the following message:
> May the gods Samas, Marduk, and Ilabrat keep you forever in good health for my sake.
> From year to year, the cloth of the (young) gentlemen here become better, but you let my clothes get worse from year to year. Indeed, you persisted (?) in making my clothes poorer and more scanty. At a time when in our house wool is used up like bread, you have made me poor clothes. The son of Adad-iddinam, whose father is only an assistant of my father, (has) two new sets of clothes [break] while you fuss even about a single set of clothes for me. In spite of the fact that you bore me and his mother only adopted him, his mother loves him, while you, you do not love me! (Oppenheim 1967, 84–85)

At this time, people viewed writing not as a means of preserving information but as a way of communicating across distance, as this son is communicating with his mother. Because of the mundane, digressive nature of so many of the documents that have been found, they have been easy to ignore. After all, what can we learn from a receipt for six sheep or the letter of a whining son to his mother? In the case of the letter, we can learn that letter writing was taught to scribes. The letters from different periods follow different formulas, but the very formulaic nature of the letters demonstrates that the art of letter writing was a skill that was taught and learned. By examining recipes for perfume, we can see that many of the principles of instructional writing that we teach today were in use in the second millennium BCE. By examining the disputations that scribal students were required to copy, we can see that the art of persuasion was a recognized skill long before Homer, the sophists, or the Greeks existed. By examining legal documents, we can understand how writing almost immediately became a tool of oppression that denied many people social justice. But do these questions justify a book-length study? Yes.

Why Do We Need This Book?

Many excellent histories of the practice and development of TPC have been written since the formation of the Association of Teachers of Technical Writing in 1973 (Tebeaux 2009). However, many phases of our discipline's rich contribution to the evolution of writing and its impact on society have not been fully explored, nor has the fact that the invention and teaching of rhetorical practices normally attributed to the Greeks can be credited to the scribes of ancient Mesopotamia, who applied them not to the act of public speaking but to the writing of transactional documents.

In this book, I seek to enlarge our understanding not only of professional communication but of the development of written communication as a whole. Those of you reading this book already know that the first form of writing was Cuneiform, but an analysis of recovered texts demonstrates a depth of rhetorical complexity not previously acknowledged. Though writing at this time was almost exclusively used for technical purposes, the writers still intentionally used persuasive techniques to increase the effectiveness of their message, and the members of this society understood the power of the written word and its ability to either grant or deny them justice.

When the tool of writing was new, those who could wield it also wielded power. Those who could not suffered the consequences. Imagine you are the only living child of a parent who has died. As such, you are your parent's *zakir shumi*. In this role, you must speak your parent's name in a ritual on the darkest night of each month. And if you don't? Your parent will haunt you, or worse, their spirit will be annihilated.

Imagine the weight of the obligation of going to your parent's grave each month for the rest of your life to chant the name that is the only thing that prevents your parent's soul from being lost to the void. Imagine your crops have failed and your family is starving. Imagine that soldiers from another city are threatening your home. You are faced with a choice. Move away and start over. Stay and meet your filial obligation and risk the lives of those you love. Imagine if the written word could save you—or doom you.

In Ur, around 2900 BCE, people began to carve the names of the dead onto funerary objects that were buried with them. "Because the phonetic signs reproduced the sounds of their name, writing had the awesome power of perpetual utterance—this is particularly credible when one realizes that at this time, reading was always done aloud" (Schmandt-Besserat 2007). Suddenly, your presence was no longer required because written words could become your voice for eternity. That is power. Imagine you were a young woman whose lover had been taken into custody by the king: "The agents of the king have seized him (my young man) in the town of Appasum, and he is being detained in the house of Nurum-lisi. But this man wears neither the fetters (of a slave) nor the hairdo of a slave. I am sending herewith Adad-sarrum to you, do send that young man back to me" (Oppenheim 1967, 82). Because the young woman had access to the service of a scribe, she could seek justice. Imagine what would have happened to the young man if she had not had such access. Imagine what might have happened to other young men who *did* wear the fetters and hairdo of a slave.

And words have even more power now because the number of people those words can reach is almost limitless. It is time to understand the ancient origins of writing and the profound and continuing effect the invention of writing had and continues to have on the world. It is time to remember, by looking back, that the words we write today have the power to define who and what we were in the past. To explore this topic, I will use the practice of cultural rhetoric in combination with other methods such as structural analysis, which I will describe next.

Methodology

As Malea Powell, Daisy Levy, Andrea Riley-Mukavetz, Marilee Brooks-Gillies, Maria Novotny, and Jennifer Fisch-Ferguson explain, the practice of cultural rhetoric doesn't require scholars to maintain the fiction that "gaps" in our history are waiting to be filled. Instead, they "believe it's important to keep all traditions/stories/histories in play as equally legitimate origins and progenitors of many simultaneous rhetorical traditions" (2014, 13). To understand our history, we must listen to the stories it has to tell, and how those stories overlap, repeat, and resonate, creating what LuMing Mao describes as hybrid rhetorics. Mao sees Chinese American rhetoric as a hybrid formed of both the tradition of European American rhetoric and Chinese rhetoric. "That is to say, that while there is no shared essence between these two traditions, there is a great deal of proximity-induced interaction, realignment, and unsettled association" (2006, 19). Further, he argues that Chinese American should be conceived of "as a process of becoming" (Mao, 19). The stories of ancient Mesopotamia were written more than 2,500 years ago. To hear them, to understand them, we must use our critical imagination.

Royster and Kirsch describe "critical imagination as an inquiry tool, a mechanism for seeing the noticed and the unnoticed, re-thinking what is there and is not there and speculating about what could be there instead" (2012, 2). Critical imagination is essential to any research that relies on the "rescue, recovery, and (re)inscription" (2012, 2) of information because it provides not a substitute for but a counterpoint to "more traditional habits of critique" (2012, 2)—habits not always useful when interpreting less easily documented forms of information such as experiences, viewpoints, and perceptions. For example, an analysis of the content of a student tablet might provide insight into the grammatical structure of the language, work that has been done by Assyriologists, but such an analysis of the ancient riddle below will not provide opportunities for imagining *why* this riddle was written:

> A house with a foundation like heaven,
> A house which like a . . . vessel has
> Been covered with linen,
> A house which like a goose stands on
> A (firm) base,
> One with eyes not opened has entered
> It,
> One with open eyes has come out of
> It.
> Its solution: the school. (Sjoberg 1975, 159)

Was this text written as a school exercise? If so, why were students being asked to write riddles, a rhetorically sophisticated form of communication. Was it meant to advertise the importance of the scribal education? Was this riddle an encomium to scribal education? Or perhaps the teacher who assigned this exercise was embracing his trickster mind (Geller et al., 2007, ch. 2) or anticipating advice that Augustine would give several thousands of years later to view play as "vital to the work of the gods," as a "divine form of subversion" (Babcock 1984, 10). Only by overlaying critical imagination onto more traditional methods can we ask such questions.

Critical imagination paves the way for strategic contemplation (Royster and Kirsch 2012, ch. 2). Strategic contemplation creates a space in which researchers listen to the hidden voices of those they study—to imagine a process of asking and answering. And again, we must stretch both our imaginations and our contemplations across the globe and through space and time. The use of critical imagination and strategic contemplation effectively combine with the practice of cultural rhetoric, which is ". . . an embodied practice. . . . Scholars must be willing to build meaningful theoretical frames from inside the particular culture in which they are situating their work. To do so means understanding a specific culture's systems, beliefs, relationships to the past, practices of meaning-making, and practices of carrying culture forward to future generations" (Bratta and Powell 2016). But can these ancient tablets, what Leo Oppenheim called "bones," tell a story? "Can documents of any kind lead a priori to reliable information about a dead civilization—especially when the texts are not intended for us? Can they guide us through the intrinsic otherness of the cultural setting that created them, and can they reveal to us a functioning 'cosmos'?" (Oppenheim 1967, 56).

Mao would answer in the affirmative because the mere fact that we were not active members of the culture that produced the Cuneiform documents I study doesn't mean we can't still understand them. About his own work, Mao says that "By characterizing [the] emergent hybrid rhetoric [that he studies] as *Chinese American*," he is not suggesting "that only Chinese Americans use and experience this rhetoric" (19). As with any rhetoric, he acknowledges that it can be used by anyone as long as Chinese and European American rhetorical traditions are being brought together and as long as relations of power continue to make their presence felt in the process (19). So, while we are not ancient Mesopotamians, we can understand the hybridization of orality and literacy, both in an ancient context and a modern one, if we remember to consider the cultural context and the relations of power and social justice in which the process of hybridization occurs. Because Mao's stipulation is so important to my project, I chose a methodology that would allow me to "build meaningful theoretical frames from inside the particular culture in which [I am] situating [my] work" (Bratta and Powell 2016). However, because I cannot read Cuneiform, to do my work I must use translated documents.

The Trouble with Translations

Working with historical documents, particularly translations, comes with its own challenges. As J. J. Connor (1993) argued, scholars must become familiar with all the published scholarship about their texts, locate them within scholarly editions, identify the texts' genres, and consider their historical context. First, technical communicators should remember that while a text might be new to them, it is not necessarily new to scholars in fields other than their own. The texts that I am analyzing have been discussed by Assyriologists who study the Cuneiform culture of the ancient Middle East. These publications laid the groundwork for my understanding of both the content and historical context of the documents and their writers. These same scholars also provided the translations on which I based my analyses, and the choice of a translation can also be problematic. According to Bellos, you can give one hundred well-known translators the same page of text, "and the chances of any two versions being identical are close to zero" (2011, 8) because translations are always approximate. Connor (1993) offers a solution to this problem: use a critical edition if possible, but when a critical edition is not available, use the best scholarly text available. I have relied heavily on the materials provided by Oxford University Press's *The Electronic Text Corpus of Sumerian Literature*. These translations are valuable because older translations are edited as new discoveries are made, and the history of each translation and the translator is made available. Though none of the other translations I used was published

Figure I.1. eme-bal, Language turner. *Source*: Bellos 2011, 29.

8 | The Origins of the Art of Professional Writing

in a critical edition, each was translated by an expert and published in a scholarly book or journal. This does not eliminate the potential for varying or incorrect translations, but the damage that would be done by not analyzing the rich trove of texts recovered outweighs concerns regarding the translations themselves.

It is also important to place the writing of a specific period within a historical context because when analyzing historical sources we must understand the time, beliefs, and the intellectual traditions of both the writer and the period in which he/she wrote (Connor and Connor 1992).

The Historical Context of Writing in Ancient Mesopotamia

Cuneiform was invented in direct response to the growing culture's need for a system of bookkeeping. In fact, the first genres of writing in ancient Mesopotamia were scribal exercises, lexical lists, and accounting documents. As the culture became more stratified and complex, the texts being produced also evolved, and new genres emerged including literature, letters, technical manuals, and legal documents—again the majority of the documents recovered and translated were examples of TPC.

Figure I.2. Disk tokens. *Source*: Musée du Louvre, Département des Antiquités orientales.

Initially, ancient Mesopotamians used clay tokens of specific shapes to represent numbers of goods. These tokens provided a method of quantifying a sale or trade of a commodity such as sheep, wool, or olive oil. In some instances, the tokens were engraved with symbols that represented the thing being counted. Eventually, scribes took these first symbols and combined them with new symbols to create the system of writing that replaced the tokens—Cuneiform (Schmandt-Besserat 1996). Cuneiform, however, was not a language. It was a method of recording the spoken language, Sumerian. Later it was used to record other major Mesopotamian languages (see table I.1).

Table I.1. Timeline of Major Mesopotamian Languages and Periods

4000 BCE Predynastic period	3200 BCE Early Dynastic Period	2000–1000 BCE Old Babylonian/ Old Assyrian Period 1000–1500 BC Middle Babylonian/ Middle Assyrian Period	1000–70 BCE Neo-Babylonian/ Neo-Assyrian Period	0 AD Late Babylonian Period
Urbanization/ use of clay tokens to count.	Sumerian is spoken language. Cuneiform is invented to record Sumerian. From 2500–2000 Old Akkadian is also spoken.	Middle Assyrian (a dialect of Akkadian) is the primary dialect spoken and begins to replace Sumerian. Enheduana writes first works of literature in Sumerian. Majority of recovered Akkadian documents are letters, legal documents, and business receipts.	Sumerian is written language of scholars. Akkadian is lingua franca. *Epic of Gilgamesh* is written.	Last known Cuneiform text is written.

Source: Created by the author.

The Uruk tablets, the oldest found (written around 3500 BC), include more than two thousand symbols; however, this number was gradually consolidated to about six hundred. Initially pictographic in nature, the symbols became more abstract, and some acquired phonetic values. Because of the complexity of the system, scribal schools quickly sprang up, which is why only three genres of writing (scribal exercises, lists, and accounting documents) initially existed (Halton and Svard 2018, 10). It was almost seven hundred years after writing was invented that the other genres developed. Sumerian, the only language spoken in the fourth millennium, faded from use in the third millennium and was used only for scholarly writing. Akkadian and then Assyrian (regional dialectics of Sumerian) became the languages of business and commerce. Most of the texts recovered from Assyria and Akkad from 2000 to 1000 BC are letters, business receipts, royal inscriptions, and legal texts. The table on page 9 outlines the different periods and languages:

Chapter Outline

I begin the book by discussing the development of the cylinder seal, a carved tube that the owner could roll over a piece of wet clay to "seal" a transaction. Cylinder seals were a technological breakthrough that enabled someone to write their "signature." Possession of a seal, or lack thereof, along with the quality of the seal itself, had repercussions for the Mesopotamians who made and used them. Today, someone might be judged by the color of their skin, or the quality of their clothing. In ancient Mesopotamia, people were judged by the quality of their seals and were even excluded from business opportunities if they didn't own one.

Several hundred years after the invention of writing, only three genres existed: scribal exercises and lexical lists, which were used for teaching, and accounting tablets, which had a professional purpose. However, seven hundred years later other genres such as religious writing, technical manuals, and letters began to appear (Halton and Svard 2018). This was a logical result Powell (2012) suggested because the writing system of any group of users develops specifically to satisfy the "needs they place upon it" (14). Consequently, it is not surprising that the two most used genres, the technical manual and the administrative letter, were forms of instructional writing.

Contrary to current scholarship in professional communication, which places the first women professional writers in the period of 1641–1700 AD, the first professional documents were written by women in 2400 BCE—eight centuries earlier. Enheduana—the first woman writer and the first nonanonymous author ever identified—wrote many of the period's great poems, including *A Hymn to Inanna*. Her work calls into question our discipline's belief that persuasive writing began with Homer and was conceptualized largely by men. This fact has the potential to completely revise the history of both professional and persuasive writing along with women's role in that history.

The teachers of the *edubba* left very little textual evidence of their teaching methods or philosophies. However, archaeologists have recovered many tablets from what they believe to be schools. By using these student texts, we can work backward from the textual evidence to reconstruct the educational process that produced them. The purpose of the edubba was to create a well-rounded writer who could work in many fields, "that is, [the edubba] was first established for the purpose of training the scribes required to satisfy the economic and administrative demands of the land, primarily those of the temple and palace" (Kramer 1981, 4).

Richard Enos argued that "rhetoric did not originate at a single moment in history. Rather, it was an evolving, developing consciousness about the relationship between thought and expression. This sensitivity about speaking, and (later) writing, happened in a variety of ways, at different times, and in several different areas of Greece" (Enos 1993, ix). This same sensitivity developed in ancient Mesopotamia thousands of years before Greek civilization existed. Although we can find evidence of rhetorical awareness in many genres of Cuneiform texts, it is most evident in the disputation literature. These dialogues between archetypal figures such as Hoe and Plow, or Fish and Bird demonstrate that what we later identify as principles of sophistic rhetoric—mythos, logos, and nomos—are present in the disputations.

In ancient Mesopotamia, people's ability to conduct business, give instruction, get a loan, avoid punishment, and own property was not based just on their gender, age, education, or social standing, but on their access to the written word. Even a slave girl could demand social justice if she could hire a scribe to write a letter for her: "What I have told you now has happened to me: For seven months this (unborn) child was in my body, but for a month now the child has been dead, and nobody wants to take care of me. May it please my master (to do something) lest I die" (Oppenheim 1967, 85). Social justice calls for collective action. Clearly, we cannot act upon a collective structure that existed thousands of years ago. However, we can take what we learn from the culture's mistakes and use it to help us identify and avoid similar errors in the future.

The line between myth, magic, and medicine was greatly blurred in ancient Mesopotamia. But that does not mean that medicine wasn't practiced. The earliest incantations come from Sumer—many of these were used to cure sickness. While these incantations remained in flux for many centuries, by the first millennium these incantations had been "canonized" for use by the mashmashu—a type of priest who was a royal official and the "principal recourse for exorcisms or for cures of illness" (Goff 1956, 5). However, rather than treating disease, these rituals were often used to punish and oppress.

Conclusion

The research in this book is not meant to be definitive. I have only scraped the surface (sorry, I couldn't resist a tablet-related pun) of the practice of technical and professional communication in ancient Mesopotamia. Many discoveries are waiting to be made, and many texts deserve to be analyzed. I do hope, however, this book will pique your curiosity and challenge some of your assumptions. We live in a world of communication technology, and it is easy to fall into the mindset that today's technology is the most "technical" ever. But we need to remember that our technology was built from the bones of the communication technologies that came first. Clay tablet, computer tablet: are they really that different in terms of their purpose and their impact? I don't think so.

Chapter 1

Cylinder Seals

Written Communication's First Technological Breakthrough

You've just purchased five sheep from your neighbor; however, those sheep will continue to live in her field until you slaughter them next month. How do you prove those sheep are yours when they are still on the seller's property? You prove your ownership by showing the receipt that both you and the seller signed with your personal signature. Of course, you don't know how to write Cuneiform and neither does she, and you don't want to pay for the services of a trained scribe. Luckily, you don't need to because you own a cylinder seal. After your purchase is complete, the seller gives you a clay envelope holding five clay tokens—one for each sheep you bought. As you and your son watch, the seller places a piece of wet clay over the opening of the envelope. Next, she takes the carved cylinder seal hanging around her neck and rolls it over the wet clay, leaving an imprint of her seal's design. She then hands the envelope to you so you can roll your own seal over the wet clay. Finally, your son adds his seal. You smile and thank her. You have just sealed your deal.

    Literally thousands of cylinder seals have been recovered from the sands of Iraq and Iran. These tiny tubes of stone, clay, and precious metals are too technically complex not to have had value to those who owned them. And the images carved on them—gods, goddesses, mystical animals, kings and queens, midwives, and weavers—clearly have much to tell us about the people who made and used them. As a seal is rolled across a piece of clay, an entire moment in history is produced; a window to the past is opened, and we can watch a king defend his people from an attacking lion. We can journey with a scribe as he travels with his prince to a foreign country (see fig. 1.1). We can admire the fashions and hairstyles worn by wealthy women at banquets. We can even ponder the reason so many seals show people drinking from straws.

But how do you make sense out of a tiny (usually one inch high and two inches wide) scene of great technical detail impressed on a piece of three-thousand-year-old clay? It's like trying to understand an entire movie by looking at one frame from the reel of a full-length film.

David J. Boje suggests we view these tiny scenes as antinarratives: antinarrative is fragmented, nonlinear, incoherent, collective, unplotted and prenarrative speculation: a bet (2001, 1). Boje developed his methods to analyze the polyphonic and collectively produced stories of modern organizations. Standard narrative analysis, Boje argues, doesn't work in this context because it pretends to transform what is anecdotal into historical fact. Antinarrative methods let us achieve several purposes (and avoid several traps). We can

- answer the question "What is going on here?"
- understand the meaning of events based upon the locality, the prior sequence of stories, and the transformation of characters in the wandering discourse
- understand the meaning as collective memory (Boje 3–4)

The scenes carved on cylinder seals are excellent examples of antinarratives. As we look at them, we can speculate about what is going on

Figure 1.1. Cylinder seal of Kalki the Scribe. *Source*: British Museum of Art, © The Trustees of the British Museum, CC BY-NC-SA 4.0.

among the people and animals depicted—and what those images signify at an ideological level, but we can never know for certain—just as we can't assume what the images represented for the person who owned the seal. Was someone who chose images that represented their craft demonstrating pride in their skill? Or even advertising their abilities? Was someone whose seal depicted their king engaged in warfare showing support for the ruling powers to benefit themselves? Did someone with a cheap, generic seal feel shame each time they used it? Did a slave who used a seal that identified her as a slave feel branded? Was that seal another type of fetter?

We do know that different seal workshops favored different scenes, and these scenes were usually tied to the local myths and nomoi. The scenes depicted, many of which are mythic, are based on the collective knowledge and memory of the people who made and used them. For example, Gilgamesh is often featured on seals. Because some seals include brief messages in Cuneiform (usually the name of the seal's owner), we can add that information. However, what we have is a tiny fragment of a larger narrative—one frame in a full-length feature film. We can't know for sure, but we can speculate. What we can know more definitely is who owned seals and how and why they used them. But what did the scenes on their personal seals mean, and why did they choose them? This is still a mystery.

Cylinder Seals

Cylinder seals first appeared in the ancient Near East in the fourth millennium, and they continued to be widely used until the end of the Persian empire in the fourth century BCE (Wiseman 1977). However, seals in various forms continued to be used for centuries. The seal used by Henry VIII (see fig. 1.2) was called "the third great seal" and was used by Henry from 1542 to 1547. The purpose of the seal, which would have been hung from a document by the attached string, was to represent the king's signature. When a seal was decommissioned, it was broken before witnesses (2022).

Henry's seal tells a very interesting story. Henry's title, "Defender of the Faith," was stripped from him by Pope Paul III in 1538. However, this seal still includes the words *Fidei Defensor* because Henry asked the English Parliament to reinstate his title, which they did in 1544 in recognition of his defense of the English church (2022).

16 | The Origins of the Art of Professional Writing

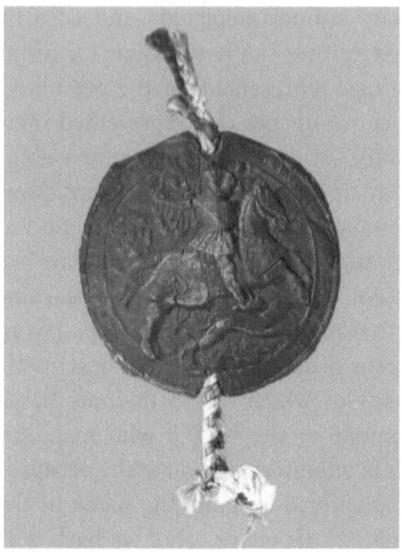

Figure 1.2. Great Seal of Henry the Eighth. *Source*: Society of Antiquaries of London, MSS/1008/A7.

Seals have a unique ability to help us understand the people who made and used them for two reasons. First, unlike most archeological artifacts, cylinder seals exist virtually outside of time. If I roll a five-thousand-year-old seal over a piece of wet clay, I will see *exactly* what the owner of the seal would have seen (see fig. 1.3). Secondly, because the seals communicate with visual images rather than the written word, we need no translation to guess at their meaning.

Figure 1.3. Contest scene. *Source*: British Museum of Art, © The Trustees of the British Museum, CC BY-NC-SA 4.0.

Consider the seal design in figure 1.3. Even as modern readers, we recognize what seems to be a battle between good and evil on this third millennium seal. The nude bearded hero in the middle wears what appears to be some sort of crown and a braided belt and is holding the bridles of two bulls. Is he protecting them from the lions? To the right a bull-man is fighting off two lions. To the left, another bull-man fights off one lion. The nude hero is a traditional figure featured on many seals from the Early Dynastic II period (2700 BCE). The two human figures have the ears and horns of a bull and bulls' bodies from the waist down. These bull-men were also a popular seal motif (Collon 1995).

We can guess at an ideological meaning for the scene—good will be tested by evil and overcome it with strength and bravery. More specifically, during this period, the separation between palace and temple grew, and kings were often featured on seals. Bulls were essential to the lives of the Sumerians—they worked in the fields, and they were eaten. A king protecting two bulls from lions, a real threat, would have represented a king doing his duty to protect his people from hunger. Yet how is it possible that we can understand a message so old?

If I were to provide a summary of this scene in Cuneiform, it would mean nothing to you without a translation. But in reality, we do not need a translation of the images. Why? Because images are universally recognized despite the boundaries of time, culture, and language (Graphic Storytelling 1996), which is why visual images are a highly effective medium for sharing technical information.

Cylinder Seals as Technical Communication

What distinguishes technical communication from other forms of communication is its laser focus on the needs of the audience. All genres of technical communication, regardless of medium, prioritize the reader by making the information communicated accessible, both in terms of its content and the form in which that content is presented (Lannon and Gurak 2017). A piece of technical communication includes only information pertinent to the reader, and it logically organizes that information so that readers can easily find what they need and then presents it in a medium that allows readers to efficiently access it. It makes use of specific linguistic strategies such as the use of imperative voice to help readers understand what the document is asking them to do.

Technical communication is not exclusively verbal. Some forms of technical communication combine words and images, and some forms of technical communication, such as cylinder seals, communicate mostly with images. And the antinarratives told on cylinder seals are "more than just another way for the field [of technical communication] to present information; [they have] significant power to engage readers, create user-centered design, [and] reach important audiences we might otherwise fail to reach" (Yu 2017, 8). Cylinder seals are a tool for communicating, and the texts they produce are an example of technical communication.

A Tool for Communicating Professional and Technical Information

As a tool, cylinder seals were intentionally designed to allow the user to easily roll the seal over a piece of wet clay to "seal" a transaction by applying what was recognized as a signature. A fitting example of human-centered design, seals often had the center drilled out and replaced with a rod that allowed the user to roll it—imagine a tiny rolling pin for pastry. Some had handles that served the same purpose. Most people wore their seals around their necks to keep them accessible. The cylinder seal is actually the first tool that allowed humans to mass produce a single image, and that image served the function of a signature before a system of writing had been invented (Teissier 1984, xxii). It might seem far-fetched to compare a strip of images to a signature, but consider that most adults' signatures bear little resemblance to their normal handwriting. In fact, many signatures are not recognizable as the words they represent (see fig. 1.4). We know their meaning by their association with the person who makes the marks and by the fact that only that person can duplicate those marks. Would you have read this as "Barack Obama" if I hadn't translated it for you?

Figure 1.4. President Barack Obama's signature. *Source*: US Government, Public domain, https://commons.wikimedia.org/w/index.php?curid=5760094

When someone rolled their seal across a piece of wet clay, they produced a strip of images—a signature. Impressions of cylinder seals were first found on *bullae* (see fig. 1.5)—clay containers that held small pieces of clay (tokens) used to count a particular commodity, such as sheep or olive oil. The phrase, "seal the deal," literally refers to the act of using a cylinder seal to legalize a transaction. This legal aspect of sealing was one of its primary uses (Teissier 1984, xxii). Seals were also used to protect entire rooms. If a storeroom held something of value, a piece of wet clay would be placed over the lock, and the owner of the room's content would seal the lock.

The intaglio images created by the seals follow a chronological order and use a system of panels, and these images reveal a great deal about the dress, beliefs, and culture of the people who owned them because the images carved on seals were not random. In fact, "There is enough general uniformity, particularly in the representation of major deities, to suggest that the seal cutters worked with a standard iconography. Different workshops naturally had idiosyncrasies, but these are generally stylistic" (Teissier 1984, xxviii). Most importantly, these images had meaning: for the owner, for those who saw them, and for the culture that produced them.

Cylinder seals, Beatrice Teissier explains, may also "be thought to belong to a minor art form (xxviii): minor, many scholars argue, because

Figure 1.5. Large clay bulla. *Source*: Metropolitan Museum of Art, Public domain.

something so small (the average seal was about an inch long) could not possibly be compared with a monolithic sculpture or wall frieze. However, Teisser argues, "In those societies where [the cylinder seal] is the primary source of images, it becomes a major pictorial art form," equal to later artworks such as Monet's paintings. Clearly, the scope of a cylinder seal is not equal to that of a painting, "but what the seal lacks in scope, it may gain in intensity, impact, clarity of design and coherence" (xxviii).

It is these very characteristics—intensity, impact, clarity of design, and coherence that make the images on cylinder seals such an effective means of communicating technical information. As this article demonstrates, the size of the images did not diminish their impact. The level of detail in the battle scenes, for example, is specific enough for the viewer to identify the types of weapons being used and to see the emotions on the combatants' faces. We can even see the patterns on their clothing. When rolled across a piece of wet clay, the revealed images are in essence a cohesive one-act play that tells a story thousands of years old.

Types of Seals and Their Uses

The people of ancient Mesopotamia recognized three types of seals: personal, official, and temple. Each type was used to communicate technical information.

Personal Seals

Personal seals provided the owner with a method for mass producing their signature for business or legal purposes. Some personal seals bore standard iconography (see figs. 1.1, and 1.3), while other more unique images were commissioned by the owner (see fig. 1.13). Later seals often have inscriptions that identify the owner of the seal (see fig. 1.7).

Official Seals

In the second millennium, royal dynastic seals often included an inscription listing the ruler's lineage. This lineage demonstrated their right to rule. Such seals were handed down from generation to generation (Teissier 1984, xxiii). These seals were the ancient equivalent of Patents of Nobility, or even DNA testing.

Temple Seals

Temple seals were given to a god in exchange for protection for the giver and their family. These seals were sometimes used in temple transactions. The images on these seals were meant to communicate the giver's requests to the god or goddess receiving them. These seals were often large—an indication that they were more symbolic.

Seal Ideology

Just as we closely associate a person's signature with that person—a signature is meant to be a written stand-in for the missing person—so did the ancient Mesopotamians associate themselves with their seals. Not only did they use them daily to conduct the ordinary business of living (buying food, selling a commodity, getting married), they believed they essentially had magical powers and saw them as a sort of talisman (Goff 1956). In fact, seals are referenced as tools in many rituals such as those meant to ward off enemies or preserve a pregnancy. And according to one text, the different stones used to make seals also had ideological significance:

- a seal of hematite (portends) that that man shall lose what he has acquired.
- a seal of lapis lazuli (portends) that he shall have power; his god shall rejoice over him.
- a seal of crystal (portends) that he shall enlarge profits; his name shall be good.
- s seal of ZA.TU.UD.AS (portends) that (until he comes) to the grave, favour upon favour shall be bestowed upon him.
- a seal of GUG (portends) that the 'persecutor' shall not be released from the body of the man. (Goff 1956, 27)

So, if a cylinder seal allowed people, even before a system of writing existed, to sign their name, how can cylinder seals be anything other than tools for communicating for technical and professional purposes?

Seals as a Tool for TPC

We sign our names to associate ourselves with a particular action or piece of communication, to say "yes, I did this. I approve of this. This is mine." Seals were uniquely suited for achieving this purpose for two reasons: their user-centered design and their ability to be mass produced.

User-Centered Design

The cylinder seal's design meets two of the most important characteristics of good design: "discoverability and understanding" (Norman 2013, 3). Humans had already discovered how to use handheld tools for many purposes, and they had begun intentionally shaping stone to meet specific purposes—the hammer stone, the wheel, the pestle. The earliest seals, which had handles attached with a dowel that made them easy to roll over clay, were intuitive to use because humans were accustomed to using round rocks as rolling pins when cooking. Later cylinder seals had an axial hole bored in them, so they could be worn on thread around the neck or on a pin. The handle, thread, and pin made the act of rolling the seal over clay more efficient. Cylinder seals were a new technology based on an already-existing one: handheld tools.

Mass Production of a Message

Seals gave people the power to mass produce their signatures while simultaneously distributing an ideological message. Just as a graphologist studies a person's signature to reveal what their handwriting says about them, we can study the images on ancient people's cylinder seals to gain insight into who they were, what they believed, and how they lived. For example, a scene illustrating a person's presentation to their god or goddess might signify a scene the person imagines playing out when they enter the afterlife. A woman who hopes to have children might choose a seal that shows a woman holding a baby (see fig. 1.6).

Because of their use as a signature, seals were a form of mass media. Stone carvers created cylinder seals and then sold them—an early form of distribution. The ideological messages on these seals were further distributed each time the owners left their seal on a document.

This system of mass media effectively provided the members of society with information they could use to understand their world. For example,

Figure 1.6. Presentation scene. *Source*: Metropolitan Museum of Art, Public domain.

scenes of ordinary activity, Collon (1987) suggests, such as weaving cloth or giving birth, were connected with the good fortune they could bring. Other imagery on cylinder seals also provides information regarding changes in fashion for over a three-thousand-year period—important information for a society that used clothing to distinguish mortals from gods, peasant from royalty, and men from women. The inscriptions provide information about who owned the seal: often a person's name and profession, less often his or her affiliations, or the occasion for the commissioning or gifting of the seal.

Anyone who could purchase a seal could own one—"even slaves) (see fig. 1.7) are attested as owners" (Collon 1987, 105). However, while a person of wealth could choose a scene and have it carved on a precious stone (see fig. 1.10) someone of lesser means would purchase a seal from a workshop "off the rack." Workshops mass produced seals on poor-quality stone and glass. "Choice, however, was probably possible even in the mass ware of the so-called Jemdet Nasr cylinders which depicted squatting pig tailed figures or animals coarsely worked with a mechanical drill" (Porada 1980, 5). Cylinder seals could also be given to someone as a token of their position. A government official, for example, might be given a seal of office. Seals could also be gifted to someone. A master might provide a slave with a seal (see fig. 1.7), or more commonly, a father might give

24 | The Origins of the Art of Professional Writing

Figure 1.7. Seal owned by slave girl. *Source*: British Museum of Art, © The Trustees of the British Museum, CC BY-NC-SA 4.0.

one to a son. Why would someone give a slave a seal? If that slave was expected to conduct business on behalf of their owner, they would need a seal to do so. Also, the seal effectively provided proof of the owner's status as slave, effectively serving as a fetter.

Specific styles developed during different eras, and these styles reflect the messages being privileged at that time.

Proto Literate Period—Before 3000 BCE

Both writing and cylinder seals begin appearing around 3300 BCE. During the first part of this period, known as the Uruk period, the carving on seals is of very high quality, and normally depicts animals across one single scene. Some seals show herds of animals and include snakes and grass. Other seals reflect more secular scenes such as a bull pulling a plough (see fig. 1.8) on which a child is seated. Some show herdsman defending their herds. Perhaps these pastoral scenes were represented as a method of demonstrating the seal owner's employment or perhaps their status.

A man who owned a bull, had land to plough, and a son to assist him, was a man to envy. Again, these images tell a story with a persuasive message—farming is a lucrative business. During the later period, Jemdet Nasr, the quality and variety of the seals decline significantly.

In this poorly carved scene (see fig. 1.9), the person on the left is a woman carrying a bundle of reeds. Reeds symbolized the goddess

Figure 1.8. Bull pulling plough. *Source*: British Museum of Art, © The Trustees of the British Museum, CC BY-NC-SA 4.0.

Figure 1.9. Woman with wheat. *Source*: British Museum of Art, © The Trustees of the British Museum, CC BY-NC-SA 4.0.

Inanna (Wolkenstein and Kramer 1983). Inanna was known as the goddess of heaven and earth, and figures in Mesopotamian literature more prominently than any other god or goddess. She was the goddess of wisdom, love, and fertility. It is also interesting to note the positions of the two figures. The Sumerian culture associated women with the left and men with the right. Consequently, a man's divine guardian will stand to his right and a woman's to her left (Stol 2016, 17–18). The Babylonians thought that a pregnant woman carried a boy on the right side of the womb and a girl on the left, which correlates with the general belief that the right is favorable, and the left is not (Stol 2016). On this seal, this belief is demonstrated by the placing of the woman on the left and the king on the right in the position of favor. Because seals from this period do not have inscriptions, we do not know who owned it, but the subject matter suggests some possible interpretations. The person making an offering is a woman. What she holds suggests that she is invoking the goddess Inanna, but she is standing before a king. Perhaps the image suggests that the king has the ability to intercede with Inanna on behalf of the owner? Perhaps a woman who longed for a child, or hoped for a successful pregnancy owned the seal?

The emphasis on scenes of daily life during this period suggests that what people valued most was the ability to simply live lives of satisfaction provided by the ability to grow ample food to eat, bear children to love, and make products to sell. The impression left by viewing these images is one of a hardworking people much like us. They worked the land, and they built businesses. And ultimately, they prospered.

Early Dynastic Period: 2900–2300

During the Early Dynastic Period numerous seals were found at the royal graves at Ur. Of particular interest is the seal of Queen Puabi.

Queen Puabi

During this period, we find more seals bearing inscriptions identifying the seal's owner and their profession. The images representing the gods and goddesses also changed. City-states began adopting patron deities, which made it necessary for seal cutters to find a way to identify the various gods and goddesses. "For this purpose the horned headdress was developed some time around 2500 BCE or shortly after" (Collon 1995, 73). Queen Puabi's seal is evidence of this shift.

Among the Ur artifacts was a seal with the inscription "Puabi, Queen" (Schmandt-Besserat 2007, 1194). In most cases, the Ur inscriptions, unlike most texts, were not inscribed on clay tablets but on vases and bowls of gold, copper, and shell, and seals of lapis lazuli—hard and durable objects that (except for copper) did not corrode nor tarnish (Schmandt-Besserat

2007, 1205). These objects clearly represented wealth and prestige, but more importantly, the security of the person's soul (Schmandt-Besserat 2007, 1209). As Denise Schmandt-Besserat explains:

> The Ur inscriptions mark a recognition of the awesome power of writing as a means to permanently capture the sounds of speech. And given the importance of preserving names in Sumer, as illustrated by the ritual of regularly pronouncing aloud those of the deceased, it is logical to assume that casting the ephemeral sounds of names into a permanent form by couching them in writing was conceived as equivalent to a perpetual utterance. The new function of writing was supplementing, or even replacing, the role of the *zakir shumi*. (1247)

By inscribing her name on a seal illustrating a banquet, Queen Puabi intended her seal to fulfill the role of the *zakir shumi* (1198), the monthly banquet at which prescribed food and drink were consumed by her descendants as they uttered her name, a method for ensuring that the deceased's spirit was not lost. Puabi, queen in her own right, no longer had to trust her place in eternity to a mortal—the written word protected her. The image may also have served another purpose. According to Virpi Oionen (2006), images are a powerful form of communication because they make frightening topics seem less threatening. Depicting a dead queen reveling in the afterlife is certainly more pleasant to contemplate than what might happen if her lost soul returned to earth (see fig. 1.10).

Figure 1.10. Seal of Queen Puabi. *Source*: British Museum of Art, © The Trustees of the British Museum, CC BY-NC-SA 4.0.

Akkadian Period: 2340–2180

Seals from this period can be categorized as early and mature. The early seals often show images of Sargon's rule. Seals from the mature period had images of Sargon's sons (Teissier 1984, 12), suggesting the primary ideological function of the scenes was to support the legitimacy of Sargon and his lineage. During the Akkadian period several changes take place. Heraldic designs become popular and feature two pairs of contestants and an inscription panel. Depending on how it was rolled, the contestants framed the inscription (or vice versa). A particular workshop, probably royal, created mirror images of the contestants.

Two recovered seals show Indian animals (elephant, rhino, crocodile) suggesting that Sargon created trade links with that region (Charpin 1995)—an accomplishment he would have asked the royal seal carvers to feature as frequently as possible.

Kalki the Scribe

Despite the preference for iconographic consistency, those who had the money could commission seals that told a personal story, such as the seal in figure 1.1. The images on this seal tell the story of an Akkadian prince and his entourage on an expedition—we see a guide carrying a pack, two officers with axes, a scribe holding a tablet, and two servants carrying the baggage. The inscription tells us that the seal belonged to a scribe named Kalki, so we must assume he is the scribe depicted. We cannot know for sure, but perhaps Kalki, or even the prince he served, specially commissioned the seal, which depicted a high point in Kalki's career: a foreign expedition. The fact that the two guides and the two servants are much smaller than the other images tells us a great deal about the professional hierarchy at this time. Kalki is depicted as equal to the prince—verification of the importance of scribes. It may be that Kalki brought back with him the rare stone on which his cylinder was carved. Again, this seal has a complex message—although he was accompanied by guides and servants, "even princes were obliged to rely on their own two feet to travel from one place to another," so at least in that sense, each member of the party was equal (Collon 1987, 158). This image also serves as a piece of useful propaganda for the members of the scribal profession. Clearly, if scribes travel in the company of royalty and commission unique seals, scribes have very high status.

Despite the discovery of unique seals such as Kalki's, most seals used one of two popular types of images: combat or presentation scenes. Seals from this period also favor episodic scenes. Scholars have offered multiple interpretations of these scenes. Because these seals were created before the great mythological works (i.e., *Gilgamesh*, which includes the story of the flood) were written, the images are thought to be visual representations of scenes from current oral tradition, and later mythological texts correlate with this interpretation. The seal cutters used their knowledge of the oral stories to create consistent images of specific heroes such as Gilgamesh and Enkidu. These images would have been widely recognized because most people were familiar with the Gilgamesh stories.

Other scholars argue that rather than depicting specific gods, the images represent cosmic or natural forces, such as the water god who has water coming out of his shoulders, or the sun god who has rays coming from his. Because the images do correspond with the content of later texts, Teissier argues that the images are the seal cutters' attempts to illustrate oral stories. And because these oral stories served to teach the population about its cultural history, these images again are communicating technical information.

Ur III: 2100–2000
(3rd Dynasty of Ur; Sumerian Renaissance, Post-Akkad)

The prosperity enjoyed under the rule of Argon was initially disrupted by the invasion of Guti tribesmen. The Sumerians, however, defeated the Guti, and Sumerian language and culture experienced a renaissance. The Neo-Sumerian seals of this period usually illustrate one of three specific types of scenes: combat between groups of three figures, presentation of a figure to a god or god-king, and libation scenes (Teissier 1984). In figure 1.11, we see a goddess presenting a worshipper to the king, who wears a cap and sits on a padded stool—standard iconography.

The inscription reads: "Ibbi-Sin the strong king, king of Ur, King of the four quarters [of the world] // Ilum-bani the overseer, son of Ili-ukin [is] your servant" (Art 2019). This inscription suggests that the worshipper is the seal's owner, Ilum-bani, and he is being presented to king Ibbi-sin, whom he serves. Again, such images serve to glorify the reigning king by emphasizing the appropriateness of being led into his glorious presence. The image also suggests that the owner of the seal had the favor of the king.

Figure 1.11. Seal of Ibbi Sin. *Source*: Metropolitan Museum of Art, Public domain.

In the early Dynastic Period (3000–2340 BCE), the scenes depicted on cylinder seals were largely contest scenes. The combatants usually included a hero, shown full-faced, bearded, and wearing a headdress of feathers and a triple belt. He is normally in the center and either engages in hand-to-hand combat with various animals or uses a knife. Many scholars believe this hero is Gilgamesh. In addition to the animals, two other heroes are often featured. The bull man on the right, who is shown full-faced, also fighting animals, is thought to be Enkidu, Gilgamesh's companion (Wiseman 1977, 25). An unknown hero is sometimes present as well. His purpose is unclear though I would argue that he might represent the owner of the seal. All three are known by their triple belts.

The Old Babylonian/Old Assyrian Period

This period was dominated by the rule of Hammurabi. At this time, seal cutters relied heavily on the use of drills and disks, resulting in seals with heavily crowded backgrounds. Presentations were still common, but additional figures now appeared: bodiless heads, dwarves, and symbols (see fig. 1.12).

The story told by this seal is very complex. A female figure wears a horned headdress that identifies her as a goddess. She is leading a supplicant before a seated god, who is holding a cup. Normally, holding

Figure 1.12. Goddess leading a worshiper to a seated deity; bull god. *Source*: Metropolitan Museum of Art, public domain.

a cup suggested a presentation of some kind to the supplicant. Behind the king are five heads: three with horns and two without. Perhaps three gods and two humans are the audience for the presentation taking place?

At a right angle to the main scene is a man in combat with a lion (God Nargul—God of pestilence, death, and the underworld) and a man in combat with a bull. Perhaps these images serve to remind the reader of the continuous battle between good and evil constantly occurring behind the scenes of our lives. Several symbols complete the scene: a crescent and a disk, a bull-altar with a conical object on its back, a monkey, and several animal heads. The crescent and disk are likely symbols for a god or goddess, and perhaps monkey and animal heads signify that the owner of the seal has traveled to foreign lands. The seal is a particularly fine example of the Anatolian type of cylinder seal, which developed during the Old Assyrian period (Art 2019).

Middle Babylonian/Assyrian

At this time, power converged in the city of Assur in the Tigris Valley. A powerful ruling family developed a regime that expanded Assyria's power and trading routes to include both Semitic and non-Semitic peoples and reached as far as the Mediterranean (Wiseman 1977). The images on the seals of this period reflect this cultural diversity. The most popular motif was the contest scene. These scenes, largely because of improved tools, were realistic representations of heroes battling animals both real and mythical: the seal below (see fig. 1.13), which the inscription identifies as the property of Ashurme. . . . Son of Amuganni, shows a bearded hero spearing a lion. Below the lion is a deer and an ostrich stands behind the lion. Its head resembles that of a griffin.

According to Wiseman, scenes such as this were representative of the battle between good and evil: "life and death, light and darkness or health and sickness" (1977, 37). The griffin-headed ostrich may signify death. In this story, the hero is shown in great detail. He has a short beard, and his hair is in a bun. A cloth protects the arm holding the javelin. The lion holds up one paw to repel the hero while using his other to prevent the deer from escaping. For the owner of the seal this story might have reassured him of the ability of good to overcome evil. Perhaps he imagined himself as the hero protecting his family from harm in the same way a reader of today might imagine being Spider-Man or Captain Marvel.

Figure 1.13. Hero fighting lion. *Source*: British Museum of Art, © The Trustees of the British Museum, CC BY-NC-SA 4.0.

Cylinder Seals and Social Justice

In theory, anyone could own a seal. However, even with less expensive seals available, and with the potential for someone to be given a seal, many people would not have had one. Access to transportation is essential in today's society. Transportation means someone can get to their job, get to the grocery store, take the children to school, and so on. Having a car is the most reliable form of transportation, though some cities have effective systems of mass transportation that can fill the need. In many ways, seals served a similar purpose for their owners because they were necessary for engaging in most transactions. However, there was no "mass transit" option available. Consequently, many people would have been excluded from many profitable activities. Seals were also used to systematically oppress certain classes of people.

This seal of Queen Puabi (see fig. 1.10) has two scenes: one on top of the other. In the top register is the inscription: "Seal of Queen Puabi." The image shows a seated woman, perhaps the queen, being served by another woman. Her hairstyle clearly shows she is a woman. A woman stands behind her chair. Opposite her, the scene is repeated with men. Again, we see women to the left and men to the right; so clearly the practice of placing women in the inferior position was still extant several hundred years later. As I said earlier, the Mesopotamians considered the left unfavorable, which clearly means that traditionally placing women on the left was not an accident. Instead, a message was being sent and perpetuated: women were less valued than men. Consequently, even though women owned seals, by using them, they were often forced to participate in their own suppression. These ideological messages were codified during King Sargon's reign.

King Sargon developed a strict seal iconography that corresponds to "precise rules regarding the dress of the figures in different combinations, their types of headdresses and their seats; the rule applies as much to court seals as to the smaller, cruder seals of private individuals with brief, two-line inscriptions, where the worshipper is generally led by a goddess in a vertically striped robe before another goddess, who is seated" (Charpin 1995, 82). Unlike the imperial style of the Akkadian period, which allowed for some artistic freedom, this period required rigid conformation as figure 1.14 suggests.

The images conform exactly to the accepted iconography. The king is seated to the right, the goddess to the left. The goddess wears

34 | The Origins of the Art of Professional Writing

Figure 1.14. Scribe. *Source*: British Museum of Art, © The Trustees of the British Museum, CC BY-NC-SA 4.0.

a flounced vertically striped skirt and leads the worshipper. The king is offering the worshipper a cup, which represents the office being given to the seal owner (Charpin 1995, 81). Based on the inscription, the image suggests the king is confirming the position of royal scribe on the seal's owner: "Ibbi-Suen, god of his country, strong king, king of Ur, king of the Four Parts: Aham-arshi, scribe, son of Babati, is your servant." So, the scribe is being presented to his king by a goddess. So, female gods were of lesser value than male kings, who were mortal, though deified. More importantly, the woman is serving the man despite her status as goddess. Finally, because women and men had to be seen wearing specific types of clothing, sexual identity was also enforced, which was another form of suppression.

Conclusion

Cylinder seals effectively provided technical information during their time, and they still provide modern readers with technical information about the ancient Near East, and the fact that the information is presented as images makes the information even more valuable.

Because a picture truly is worth a thousand words, images are the most appropriate medium imaginable for sharing technical information about the culture that created them. The images on ancient cylinder seals allow us to literally "see" what life was like in ancient Mesopotamia. Texts do exist from this period, but a lexical list of items of clothing, or types of

Figure 1.15. Winged hero. *Source*: Pierpont Morgan Library. Morgan Seal 747.

crops, will never provide us with the same sort of visceral understanding that an image will. The images on cylinder seals give us insight into many aspects of ancient life. Images of men plowing with "seed plows" make it clear that the people living around 3000 BCE had developed and were using advanced agricultural equipment. Seals that show women weaving and spinning demonstrate that the textile industry was an important piece of the culture's economy. A seal belonging to a doctor who lived around 2000 BCE illustrates the tools he used for delivering babies. Images of warriors are so detailed that historians have been able to accurately determine the evolution of the weapon. Without these images we would have no method, other than using our imaginations, of truly knowing how people lived, worked, and worshipped at the dawn of communication. We would not know how they grew crops. We would not know how they dressed. We would not know how they envisioned their gods and kings.

Cylinder seals were the first form of mass communication. They allowed people of all classes to legally identify themselves so that they could conduct business and provide legal protection for themselves. The images they selected for their seals told us what they valued, what they believed, what they did, and where they had been. And, like all images, they communicated across boundaries. They did not privilege one language. They did not require literacy. They were not denied to anyone with eyes to see them. Imagine historians two thousand years from now trying to piece together a picture of our lives. Which will tell them more, political cartoons or political tweets?

Chapter 2

Ancient Technical Manuals and Letters

The Origins of Instructional Writing

You are a beer maker of great renown, but you are old, and soon you will no longer be able to work. What will happen to your knowledge? How will it be carried on? If your words are inscribed into a manual, then your recipes will live on even after you are gone, and your skill will be known for eternity.

Imagine your brother lives in another settlement several miles from you. Because you are the elder, he is supposed to send you barley and silver each month to help care for your elderly mother. However, your village is being threatened by thieves, and you do not want him to send you anything until you and your family are safe from robbery. What can you do? You can't leave your family and travel across many dangerous miles to tell your brother what has happened. You have no other family to send. You hire a scribe and have him "write" a letter to your brother. After transcribing your words and using his skill to make the message as persuasive as possible, the scribe delivers the letter to your brother. To your relief, the same scribe returns seven days later with a letter of response. Your brother will not send barley and silver until you command him to do so. Praise the gods.

Letters including instructions were one of the most common types of writing in ancient Mesopotamia. A logical conclusion Malea Powell suggested because the writing system of any group of users develops specifically to satisfy the "needs they place upon it" (2012, 14), and a close analysis of their documents reveals that two commonly used genres, the technical manual, and the administrative letter, can be categorized as types of instructional writing:

- The Technical Manual—Scholars have identified the technical manual as a genre of Cuneiform text "unified by a common linguistic register, characterized by conditional clauses and the use of second person verbs that instruct an anonymous doer to make a certain product or reproduce a particular skill" (Halton and Svard 2018, 220). Archaeologists have recovered technical manuals on topics such as horse training, brewing beer, making perfumes, and making colored glass. And canonized medical manuals describe how to prepare plants, diagnose illnesses, and treat illnesses. The primary purpose of these manuals, which were written from 1900 to 539 BCE, was to provide expert-level step-by-step instructions (Halton and Svard 2018).

- The Letter—Letters, Oppenheim (1964, 73) explained, were not used just for personal communication. They were also commonly used for giving administrative instructions. While these letters naturally had a smaller readership than the technical manuals, they were one of the most widely used genres for communicating instructions from one person to another across distances, making these letters an important window into the culture. According to Oppenheim (1964) although the salutations varied from one period to another, the writers of administrative letters consistently used imperative mood.

An analysis of these ancient instructional documents reveals two important points:

- Though they were written when writing as a tool was in its infancy, the writers used many of the same linguistic structures and techniques that we use today.

- Our discipline largely supports the opinion that the art of rhetoric developed in ancient Greece in the fourth century BCE, though Enos argued that the origins of the art form can be found in the eighth century in the works of Homer;

however, an analysis of excerpts from ancient technical manuals and letters demonstrates that the writers of those documents practiced many of the principles of persuasion later conceptualized by the Greeks, and they intentionally made choices designed to appeal to their audiences rhetorically.

The people of the ancient Near East, inventors of writing, fully understood that providing instructions was a highly persuasive and reader-centric act that required the writer to make specific choices—the same choices we still make today. In fact, when we write instructions and teach others to write them, we are practicing principles used by the Mesopotamians.

I will first analyze specific texts to demonstrate the writers' use of linguistic structures associated with instructional writing. The writers of both technical manuals and administrative letters used the same linguistic structures and strategies (see table 2.1). I will also analyze the rhetorical motive behind the choices each writer made.

As Brockman (1998) has shown, effective writers sometimes intentionally violate the norms of a genre to persuasively communicate with that audience. Brockman noted that while nineteenth-century women readers of sewing machine manuals were willing to be placed in the subservient position required of instruction manuals that demanded they "read from beginning to end" (224), men were not. Male readers of automobile instructions did not want to be directed; so to prevent the male reader from turning his back and refusing to read, writers of these manuals had to diminish the authority in the rhetoric (225). Simply put, women were comfortable with the use of imperative voice, while men were not. Consequently, the Ford company adapted its writing style: Ford no longer demanded a reader's attention. In fact, the author stated that, whatever else, reading the manual and thoroughly understanding the information were " 'not imperative,' but only 'desirable' " (226). Although history hasn't provided us with any reliable textual evidence to demonstrate that Mesopotamian writers were cognizant of the concept of rhetorical effectiveness, my examination of the documents presented in this chapter clearly shows that the writers were making intentional choices based on their audience—even when this meant not following the standard format for a particular genre of document—the same phenomenon observed by Brockman. I will begin by analyzing an excerpt from the technical manual, *A Perfume-making Recipe of Tapputi-belet-ekallim*.

Table 2.1. Linguistic Structure of Instructions

Use of conditional clauses	Instructions from manuals regularly use clauses such as "If you want to process aromatic cane oil" (Halton and Svard 2018, 222)
Appropriate level of detail and technicality for audience	The majority of the texts I analyze demonstrate that the writer had a strong sense of their audience, as evidenced by the level of detail and explanatory information included: "It should be full (of beads) and should be beautiful" (Oppenheim 1967, 87). Writers also separated commands from feedback statements (Halton and Svard 2018) (Oppenheim 1967)
Use of chronologically ordered steps	Although ancient writers did not number each step as we do today, the chronological order was indicated by the order of clauses, and the use of spatial prepositions, ordinal words, and verb tense.
Use of imperative voice	Within these steps, we find evidence of writers' attempts to include one task per clause, stated in imperative mood (Halton and Svard 2018): "Seal it (in a package) and give it to him" (Oppenheim 1967, 87). The *Chicago Manual of Style* defines imperative voice as "commands {**go** away!}, direct requests {**bring** the tray in here}, and, sometimes, permission {**come** in!}. It is simply the verb's stem used to make a command, a request, an exclamation, or the like {**put** it here!} {**give** me a clue} {**help**!}. The subject of the verb, *you*, is understood even though the sentence might include a direct address {**give** me the magazine} {Cindy, **take** good care of yourself [*Cindy* is a direct address, not the subject]}. Use the imperative mood cautiously: in some contexts, it could be too blunt or unintentionally rude. You can soften the imperative by using a word such as *please* {please **stop** at the store}. If that isn't satisfactory, you might recast the sentence in the indicative {**will you stop** at the store, please?}."

Source: Created by the author.

A Perfume-making Recipe of Tapputi-belet-ekallim

Tapputi-belet-ekallim, a woman, wrote the technical manual *A Perfume-making Recipe of Tapputi-belet-ekallim*, in roughly 1400–1000 BCE. A colophon on one of the tablets identifies Tapputi-belet-ekallim as the "writer" of the manual, but different translations have raised questions regarding her authorship. A 1968 translation of the colophon used the words "copied on the command of Tapputi-belet-ekallim," (Hunger 1968) while the most recent translation by Escobar uses the words "according to the mouth" (Escobar, forthcoming). Halton and Svard argue in favor of the second translation because the text of the tablet itself describes Tapputi-belet-ekallim as "having dictated the recipe, which was then written on a clay tablet and baked by a scribe" (Halton and Svard 2018, 220). Both translations suggest that Tapputi-belet-ekallim, while the source of the words, did not physically place them on the tablet. Unfortunately, we have no way of knowing more about the writer than what we can infer from her colophon and the content of the instructions themselves: "Colophon: Perfume-making recipe for 2 seahs of processed cane oil, fit for a king, according to the mouth of Taputti-belet-ekallim, the perfume-maker; according to the month *Muhur-ilani* on the 20th day; the eponymate of Sunu-qardu, the chief cupbearer" (234). Most texts at this time did not identify the writer, which suggests that Tapputi-belet-ekallim was a recognized expert whose name would have increased the reputation of the text itself—the use of an ethical appeal. The qualitative description of the product being made, "fit for a king," is another example of ethos. She clearly had established expertise and access to people at the highest levels. Finally, based on the technical level of the content, we must assume that Tapputi's targeted readers were familiar, at least to some extent, with the process of making perfumes.

The following translation uses parentheses to "ease the flow of the prose; reconstructed or broken sections are represented by square brackets" (Halton and Svard 2018, 222), so I am not including words in parentheses in my analysis. Excerpts from the manual follow:

Obverse Column 1

Line 1

If you want to process aromatic cane oil: (take) 2 seahs worth of cane, along with their *tubaqu*-roots (i.e., the whole cane).

Lines 2–4

Once you have washed them, you set down a *sahistu*-sized *diqaru*-vessel and heat *tabilu*-aromatics with fresh high-quality water from a palace well of Assur. You transfer (the mixture) into a *haru*-vessel.

Lines 4–6

You (then) pour on top of this liquid mixture, within the *haru*-vessel: 1 liter of *hamimu*-aromatic, 1 liter of *jaruttu*-aromatic, (and) 1 liter of myrtle, good-quality (and) filtered.

Lines 6–8

These are your measurements, to be apportioned according to the amount of water taken. You perform (the steps prescribed) at sunset and at nightfall! (The mixture) is to steep overnight.

Lines 8–10

At dawn, when the sun rises, you filter the liquid and these aromatics through a *sunu*-cloth into a *hirsu*-vat. You clarify the mixture (by filtering it) from this *hirsu*-vat to another *hirsu* vat.

Lines 11–12

You remove the *minduhru*-particulates. You wash 3 liters of crushed nut-sedge with the liquid mixture of these aromatics. You remove the *pahhu*-particulates.

Lines 13–14

You put on top of this liquid mixture of aromatics, within a *hirsu*-bowl: 3 liters of myrtle, 3 liters of cane, crushed and filtered.

Lines 14–16

You measure out 4 seahs of this liquid mixture that has (steeped) overnight with aromatics.

Lines 16–17

You filter through a sieve: 1½ liters of unfiltered mash made from almonds (together with) 2 cup-fulls—(using) small cups—of wood shavings from the *kanaktu*-tree.

Lines 17–19

You gather up the oil (produced) in a *haru*-vessel. In the liquid mixture . . . (222–23)

Use of Linguistic Techniques and Structure

Even to modern readers, *A Perfume-making Recipe of Tapputi-belet-ekallim* feels familiar because it follows the standard conventions of a well-known genre. Anyone who had technical knowledge of the tools and supplies used in the recipe could probably follow her instructions successfully because of their use of linguistic techniques and structures intended to provide the reader with needed information in the appropriate format.

USE OF CONDITIONAL CLAUSES

Conditional clauses often begin with the conjunction "if," though many other conjunctions can be used to create conditional clauses[1] and normally precede the independent clauses they modify. They state a hypothesis or condition. The excerpt that I included uses two conditional clauses. The first clause begins the recipe: "If you want to process aromatic oil" (222). The purpose of this conditional clause is simple. It asks the reader: Do you want to make perfume? Logically, this device allows readers to decide, based on what they have already read, whether they want to continue following the instructions by ensuring that they understand what the outcome will be. Modern instructions also begin by providing readers with the information they need to decide whether to follow a set of instructions: What do I need? How long will it take? What skills are required? What will the outcome be?

The second conditional clause is in line two: "Once you have washed them, you set down a sahitsu-sized diqaru vessel." In this example, the conditional conjunction "once" suggests that the second act of setting down the vessel is conditional upon the first act of washing the roots named in the previous line. This use of a conditional clause is significant because it demonstrates that the writer understood the need to make the chronological order of the steps clear. Because writers at this time did not number each step, such linguistic constructions ensured the readers' understanding of the proper order of the tasks being completed.

Use of Imperative Mood

Constructions such as "you put," or "you measure," could be considered indicative rather than imperative. However, I believe that considering the purpose of Mesopotamian technical manuals within the context of the culture that produced them allows for a strong argument to be made for reading these constructions as examples of imperative voice that include the normally understood "you," which as Halton and Svard stated, is the norm for technical documents: "As a genre, technical manuals are unified by a common linguistic register characterized by . . . the use of second-person verbs that instruct an anonymous doer ('you') to make a certain product or reproduce a particular expert skill" (220). As Halton and Svard specify, recipes such as the one I include here were used to "instruct." The writer's intention was to share her expert knowledge in the form of step-by-step instruction. Her use of the second person followed by a second-person verb serves to instruct the reader to complete the specified task. Despite the use of "you," which is normally understood, I believe that based on the writer's intention, we can categorize this linguistic structure as imperative voice. The *Chicago Manual of Style* supports my categorization:

> The imperative mood expresses commands {**go** away!}, direct requests {**bring** the tray in here}, and, sometimes, permission {**come** in!}. It is simply the verb's stem used to make a command, a request, an exclamation, or the like {**put** it here!} {**give** me a clue} {**help**!}. The subject of the verb, *you*, is understood even though the sentence might include a direct address {**give** me the magazine} {Cindy, **take** good care of yourself [*Cindy* is a direct address, not the subject]} (5: Grammar and Usage 5.122: Imperative mood 2020)

In the case of the text being analyzed, "you" would replace "Cindy" in the example above.

The writer uses imperative mood thirteen times.[2] The purpose of imperative mood is to give a command. In modern English, the subject "you" is normally understood. However, the writer of this technical manual chose to include "you" before the imperative verb: the only clause that does not use imperative mood is the opening conditional clause discussed above. The writer also disrupts the use of imperative mood in line six: "These are your measurements, to be apportioned according to the amount of water taken" (222). This line serves as a transition from the first section of the recipe to the next. It tells the readers that they have now completed assembling their ingredients and can determine specific amounts based on the amount of water they choose to use. Her use of passive voice—"to be apportioned according to the amount taken"—seems intentional. Because she is writing to an audience of experts, it does not seem persuasive to command them to do what is obvious.

Appropriate Level of Detail for Audience

Whether the writer used an appropriate level of detail is more difficult to determine because we cannot know with certainty the intended audience for instructions written in approximately 1400 BCE. However, contextual clues suggest that Tapputi-belet-ekallim was writing for a specific audience who had specialized knowledge. For example, she refers to water from a particular location: fresh high-quality water from a palace well of Assur. Clearly, she assumed that her readers were familiar with this source and had access to it. Her failure to describe or define the specific ingredients and measurements also suggests that they were familiar to her audience. I am inclined to believe that the writer made a rhetorical move when she assumed her audience did not need this information. The colophon uses the words "fit for a king" when describing the perfume that the reader can make if he or she follows the words of Tapputi-belet-ekallim. Why would she risk her reputation by failing to provide sufficient detail for her readers when she was careful to do so elsewhere? She does, however, provide detail in many cases:

- Specific measurements—The writer provides eleven measurements: 2 seahs worth of cane / a *sahtisu*-sized vessel / 1 liter of *hamimu*-aromatic/ 1 liter of *jaruttu*-aromatic / 1 liter of myrtle / 3 liters of crushed nut-sedge / 3 liters of myrtle / 3 liters of cane / 4 seahs of liquid / 1½ liters of mash / 2 cups-full

- Specific times—at sunset and at nightfall! / overnight / At dawn, when the sun rises / overnight

- Specific sources—Palace well at Assur
- Specific qualities—fresh, high-quality water / good quality / small cups
- Specific tools—*Diqaru*-vessel / *haru*-vessel / *sunu*-cloth / *hirsu*-vat / sieve

We can assume that her readers were familiar with these terms because she specified their use but did not describe them.

Use of Logically Ordered Steps

This set of instructions undeniably uses logically ordered steps. The writer lists the tasks to be completed chronologically, and she breaks the text into separate lines. The chronological order is reinforced by references to specific times, and the steps roughly correlate with the number of lines (see table 2.2). In lines 2–4, she uses present perfect tense rather than present tense: "Once you have washed them." However, because this dependent clause modifies the independent clause that follows, "you

Table 2.2. Number of Tasks Per Line

Line Number	Imbedded Commands
Lines 2–4	set down heat transfer
Lines 4–6	pour
Lines 6–8	perform
Lines 8–10	filter clarify
Lines 11–12	remove wash remove
Lines 13–14	put
Lines 14-16	measure
Lines 16–17	filter
Lines 17–19	gather
19 Total Lines	**15 total Commands**

Source: Created by the author.

set down a . . ." the writer used present tense to clarify the sequence of actions: do this, then do that.

Rhetorical Moves

The writer's consistent use of expected linguistic structures suggests that she was familiar with the genre of the technical manual. However, some stylistic deviations seem to have rhetorical motives.

Use of Passive Mood

The writer veers from convention by using passive voice on one specific occasion: "These are your measurements, **to be apportioned** (emphasis mine) according to the amount of water taken." Why did she use passive? Perhaps she felt it was inappropriate to command her readers to do something that seems like common sense: apportion your ingredients based on the amount of water *you* choose to use. Or she might have chosen to use passive voice because she cannot know how much water her readers might choose to use—and if she doesn't know, how can she give a command? Regardless of which interpretation we choose, and given the age of the recipe no choice can be honestly justified, her singular use of passive voice suggests that she made an intentional rhetorical decision based on her audience's needs. Letters use the same linguistic devices but add a higher rhetorical sophistication to "render the audience receptive" to the instructions being sent.

Letters

Today, we often describe the structure of a letter using the terms "direct" or "indirect," and neither structure normally uses imperative mood. However, most letters written in ancient Mesopotamia did use imperative mood. According to Oppenheim: "The style of a letter or message took one of two forms (Oppenheim 1964, 222) based on the period in which it was written.

Sumerian to Neo-Babylonian (3000–600 BCE)

The form of letter that developed during this period was used for giving administrative instructions and began with the order to "recite the message verbatim (Say to PN . . .)" to the addressee named in the heading of the letter. The instructions are almost exclusively given in the imperative and concern administrative matters, "normally the delivery of goods or animals" (Oppenheim 1964, 277): "Tell Ina: Ikuppija, Ellil-bani, and Assur-taklaku send the following message" (Oppenheim 1967, 73).

OLD BABYLONIAN (2000–1600 BCE)

The Old Babylonian form that developed later used "more or less elaborate blessings and greetings inserted after the heading—according to the social relationship between writer and address" (Oppenheim 1964, 277). Most blessings invoked the name of a god or goddess:

> "May the gods Samas and Marduk keep you forever in good health." (Oppenheim 1967, 78)

> "May the gods Samas, Marduk, and Ilabrat keep you forever in good health for my sake." (84)

Although both forms were in use simultaneously, the formulaic blessings do not appear before the Old Babylonian period.

Though writers (writer means both/either the speaker of the message and the scribe who transcribed it) did adhere to the expected stylistic conventions of each form, as Oppenheim stated, writers still had ". . . considerable freedom in presenting [their] case to the addressee, shifting from argument to argument,[3] changing topics, returning to previous points, in short, making the fullest use of the language as an instrument to convey a complex message" (64–65). However, as Oppenheim cautioned, we must consider the gap that exists between the intended message and the wording. Because most letters were dictated to scribes, the scribe's training, and social position in relation to the person speaking the message would have affected the words chosen. Some scribes were highly trained and had the social position to actively collaborate in the writing of a letter, while others were "poorly paid town scribe[s] who translate[d] the inarticulate complaints of the poor and uneducated into the stereotyped eloquence of a petition or begging letter" (65). Because I am looking at the use of standard stylistic conventions, whether those conventions were met by the speaker or the scribe does not change the fact that the conventions were known and met. The next letter is from the Old Babylonian period.

"Give the Silver to Su-Belim"—Letter from a Sister to her Brothers (emphasis mine)

First translation. From Textes cuneiformes du Louvre.

> Tell Assur-risi, Su-Belim, and Assur-taklaku: Elani sends the following message:
> Dear Brothers, **get hold** of Isme-[. . .] and Assur-nada there: **make them pay** nine shekels of silver, and **give** the silver to Su-Belim, **Be sure**, dear brothers, to give the silver to Su-Belim, lest you cause annoyance to me and to him. When I stayed there, they (the two debtors mentioned) told me the following: "No sale can be made on the market." Today I hear, however, that many sales are being made on the market. Therefore, **make them pay** the silver and **give** it to Su-Belim. **Follow** instructions, dear brothers! (77)

This letter relies heavily on the use of specific linguistic structures.

USE OF ADVERSATIVE CLAUSE

The writer uses one adversative clause: "lest you cause annoyance to me and to him," "him" referring to one of her brothers, Su-Belim. Clearly, the contrast being made is between what might happen if the brothers do or do not obey their sister. She is making an implied threat regarding the danger of angering herself and Su-Belim. Elani is giving the orders to her three brothers. She is in charge. However, she has given Su-Belim authority over her other two brothers, Assur-risi and Assur-taklaku. By using only one conditional clause, Elani strengthens the impact of her implied threat. However, naïve overuse of conditional clauses dilutes their effectiveness.

USE OF SECOND PERSON

In this short letter, Elani uses second-person imperative eight times.

1. **Tell** Assur-risi, Su-Belim, and Assur-taklaku
2. **Get hold** of Isme-[. . .] and Assur-nada
3. **Make them pay**

4. **Give** the silver
5. **Be sure**
6. **Make them pay**
7. **Give it**
8. **Follow instructions**

Her commands are clear and precise, and she uses repetition to reinforce what she considers most important: make them pay.

APPROPRIATE LEVEL OF DETAIL AND TECHNICALITY

The writer clearly stipulates why the readers should follow the instructions: the two debtors had refused to pay her when she visited because they said no sales could be made. However, this information is contradicted by another source, and the contradiction provides the justification for collecting the debt. By providing this justification, Elani achieves two goals. She does her brothers the courtesy of providing a rationale for the orders she is giving, and she neatly precludes any objections they might have made based on lack of justification. The reader is also told what hazards exist: if the brothers do not collect the silver, they will make both the writer and Su-Belim "annoyed." As Elani and Su-Belim control the silver, we must assume that the brothers will take this warning seriously. She also includes a level of specificity that would make it difficult for anyone to fail to do as she commands:

- Use of 6 names—writer, three brothers, two debtors
- Use of amounts—9 shekels of silver
- Use of direct quote—"No sale can be made on the market."
- Use of repetition—Give/ give / make them pay/ make them pay

CHRONOLOGICAL STEPS

Although she does not list her commands line by line, Elani does give her commands in a clear chronological order as the previous numbered list demonstrates.

Ancient Technical Manuals and Letters | 51

Rhetorical Motives

On the surface, the direct tone of Elani's letter, given the fact that her audience is her brothers, might seem inappropriate. However, she is following standard conventions for giving administrative orders, and she does use rhetorical devices to soften her tone.

USE OF INFINITIVE

Elani breaks away from her use of imperative voice to repeat the importance of giving the silver to Su-Belim: **give** the silver to Su-Belim. **Be sure,** dear brothers, to give the silver to Su-Belim. The grammatical structure of this sentence is interesting.

> Give (verb) the silver (direct objective) to Su-Belim (indirect object)
>
> Be sure (verb) to give (infinitive verb) the silver (direct object of give) to Su-Belim (indirect object of "to give")

Elani could have simply repeated the first clause; instead, she replaced the imperative construction of the first sentence, "give," with the command to "be sure." She again uses "silver as the direct object of the verb, though in this case "silver" is the object of the infinitive form of "give" rather than the imperative form. Simply put, her primary message was "give the silver, be sure to give the silver." In both clauses Su-Belim is the indirect object. This grammatical structure suggests that Elani's brothers Assur-risi and Assur-taklaku must do the giving and the ensuring of that giving—not Su-Belim. Again, Elani has subtly subjugated the power of both Assur-risi and Assur-taklaku to Su-Belim and Elani, though she does not say this directly—a rhetorical move likely to soften her command.

USE OF AN ADVERSATIVE CLAUSE

Elani's use of an adversative clause reinforces the positions of power within the family: "lest you cause annoyance to me and to him" (77). The threat, thinly veiled as a contrastive element, implies there will be consequences should her two brothers fail to follow her instructions. We have no way of understanding the family dynamics, but the letter does suggest that two of her brothers are less responsible. Why? We will never know.

USE OF PASSIVE VOICE

It is interesting that the writer uses a passive verb construction when quoting the two men who owe her the silver: "No sale can be made . . ." Given her otherwise consistent use of imperative voice, we must assume her choice to use a passive construction was intentional. Perhaps she is quoting the speaker directly, and he meant to lay blame for his failure to pay on the "market." The writer also uses an interesting mix of active and passive construction to draw attention to the most important point in her letter:

> Today I hear (active), however, that many sales are being made (passive) on the market.

Based on the sophistication of this letter's construction, it seems likely that the writer intentionally mixed active and passive voice in one sentence. The use of the active voice "hear" adds authority and immediacy to her claim, while the use of passive voice ("are being made") allows her to avoid identifying the source of her information. Perhaps the writer had a reason for choosing not to implicate the source of this information, or perhaps she did not know the name of a specific person making sales—we have no way of knowing. However, her use of passive voice, whether intentional or not, demonstrates its use as a tool of persuasion.

USE OF ADJECTIVES

The tone of Elani's letter can only be described as direct. She does, however, use adjectives to soften her tone. Elani uses the noun "brothers" three times, and each time she adds the adjective "dear." This show of filial affection was certainly an intentional attempt to gain their cooperation.

CALL TO ACTION

Should her readers have any doubt as to the importance of following her instructions, Elani ends her letter with a call to action: **follow** instructions, dear brothers! (77). This call to action is the last sentence in the letter, clear evidence of the importance Elani places on this command.

The next letter, while similar in many ways to Elani's, was written by a student who still had much to learn about how to write a persuasive letter.

Letter from Son to Father—Instructions for Ensuring Filial Loyalty

Previous translation—Goetze, *Sumer* 14, 73–74. Second translation by Oppenheim.

This letter is an example of the second form of letter, which uses a more elaborate greeting. The writer uses a standard blessing, though given the relationship of the writer to the reader (son to father), we might expect a more elaborate and personal greeting, particularly given the nature of the demands the son makes: "May the gods Samas and Wer keep you forever in good health" (87). However, all but one of the six letters included by Oppenheim in this section use some variation of the same greeting: may the god(s) _____ keep you in good health, so clearly this was a formulaic greeting such as dear _____. Other features of the letter suggest that Adad-abum understood, and was attempting to follow, the correct format for a letter. However, his failure to practice rhetoric, makes his letter much less persuasive than that of the previous writer:

> Tell Uzalum: Your son Adad-abum sends the following message:
> May the gods Samas and Wer keep you forever in good health.
> I have never before written to you for something precious I wanted. But if you want to be like a father to me, get me a fine string full of beads, to be worn about the head. Seal it with your seal and give it to the carrier of this tablet so that he can bring it to me. If you have none at hand, dig it out of the ground wherever (such objects) are (found) and send it to me. In this I will see whether you love me as a real father does. Of course, establish its price for me, write it down, and send me the tablet. The young man who is coming to you must not see the string of beads. Seal it (in a package) and give it to him. He must not see the string, the one to be worn around the head, which you are sending. It should be full (of

beads) and should be beautiful. If I see it and dislike(?) it, I shall send it back!

Also send the cloak, of which I spoke to you. (Oppenheim 1967, 87)

This writer does use the genre-specific elements of the letter; however, his letter fails for many reasons. Perhaps he is an inexperienced writer, or perhaps he is relying on his relationship with his father rather than the quality of his writing to render his message persuasive.

CONDITIONAL CLAUSES

In this letter, the son uses three conditional clauses:

- If you want to be like a father to me
- If you have none at hand
- If I see it and dislike (?) it

The writer (very) obviously uses two of these clauses not to state a hypothesis but to state a condition—if you do not do what I order, then I will punish you. He is using the conditional clause to make threats. Perhaps the writer intended the conditional nature of the clause to soften his threats. The other conditional clause "if you have none at hand," does state a hypothesis. If you can't find what I want, then make it yourself! If Elani's careful use of one conditional clause helped her to identify her priority, this writer's overuse of conditions has the opposite effect—he sounds like a spoiled child.

USE OF SECOND PERSON

The writer uses both second person and third person throughout his letter. Each time he tells his father what to do, he uses second person. However, when referring to what he will do, or what he doesn't want the "young man" to do (emphasis mine), he uses third person:

1. Get me a string of beads
2. Seal it with your seal

Ancient Technical Manuals and Letters | 55

3. Give it to the carrier of this tablet
4. Dig it out of the ground
5. Send it to me
6. Establish its price
7. Write it down
8. Send me the tablet
9. **The young man who is coming to you must not see the string of beads**
10. Seal it in a package
11. Give it to him
12. **He must not see the string, the one to be worn around the head, which you are sending**
13. Send the cloak

Why did the writer choose to use third person for steps nine and twelve, and why only when he was referring to the young man whom we must assume was the recipient of the gift? He could have used second person as he did elsewhere—do not show the young man the string of beads. And why did he repeat this twice? I believe that the writer was attempting to create some separation between his father and his friend. Why? It seems likely that the son does not want his father to know that he is requesting this expensive gift for someone other than himself. Again, we might ask why, but that is an answer history will never reveal. These same inconsistencies create confusion regarding the order of the tasks to be completed.

STEP-BY-STEP DIRECTIONS

The writer does provide a clear list of instructions in imperative voice. However, the writer sometimes disrupts the logical order of his commands by introducing conditional clauses, or additional instructions that are in active voice but use third person rather than second.

Part 1—Prove You Are Like a Father to Me

If you want to be like a father to me (conditional)

1. Get me a string of beads
2. Seal it with your seal
3. Give it to the carrier of this tablet

Part 2—If You Can't Buy the Beads, Make Them

If you have none (conditional)

1. Dig it out of the ground
2. Send it to me

Part 3—Prove You Are like a Father to Me by Proving the Bead's Value

In this I will see whether you love me as a real father does (conditional)

1. Establish its price
2. Write it down
3. Send me the tablet

Part 4—Hide the Beads

The young man who is coming to you must not see the string of beads—third person

1. Seal it in a package
2. Give it to him

Part 5—Hide the Beads

He must not see the string—active, the one to be worn around the head, which you are sending—third person

Part 6—Make Them Exceptional

It should be full of beads and should be beautiful—third person

Part 7—Do What I Say, Or Else

If I see it and dislike it (conditional), I shall send it back—first person, active

Part 8—By the Way, Send My Winter Coat

1. Send the cloak

This order is disruptive and confusing for several reasons. First part 3, which asks the father to legally value the beads, follows the previous commands to either give the beads to the messenger or send them to the son. How can the father value beads he has already sent? Also, wouldn't the description that the son provides in part 6 have been more useful at the beginning?

Rhetorical Motives

The rhetorical implications of the writer's choices are even more intriguing.

APPROPRIATE LEVEL OF DETAIL AND TECHNICALITY
FOR AUDIENCE

The son makes many assumptions regarding his father's knowledge of jewelry. We have no way of knowing if the writer's father was a jeweler; if he were, then he might have the specialized knowledge needed to know how to make beautiful beads out of what he can dig out of the ground and then fashion those beads into a full and beautiful string for the head. He might also know how to appraise the value of those beads. However, subjective descriptions make the father's job very difficult. How many beads is full? Based on the demanding tone of the letter, and the son's apparent ignorance regarding the crafting of a string of beads, I find it more likely that the son was simply demanding what he feels his father owes him. We can only hope the son received neither his cloak nor the string of beads. Perhaps such a lesson would have encouraged him to become a more persuasive writer, not to mention a more considerate son.

USE OF PASSIVE VOICE AND THIRD PERSON

The writer uses passive voice in one phrase that is repeated at both the beginning and end of the letter: "to be worn about the head." The writer never states who the beads are for, but he does provide a strong clue in this sentence: "He [the person collecting the beads] must not see the string of beads." To prevent the messenger from seeing it, the son commands his father to seal the beads in a package. It seems possible that this person, also identified as "the carrier," and the "young man" is the intended recipient of the gift. Although the son may feel that his use of passive voice will prevent his father from understanding his motives, it does not seem likely.

OPENING APOLOGY

Although the writer of this letter predated Aristotle, he still attempted to begin his letter by rendering his audience receptive: "I have never before written to you for something precious I wanted" (Oppenheim 1967, 87). Sadly, his justification is more of a threat. Clearly, the writer of this letter did not understand the importance of appealing to his audience.

This letter is a clear example of instructional writing—poorly written instructions. The writer used a demanding tone; he did not provide adequate information, the steps are not logically ordered, and he did not separate feedback from tasks. These instructions are an excellent example of what not to do when writing instructions.

USE OF THREATS

As I stated earlier, not only is the sequence of the instructions flawed, but the writer often interrupts his instructions to make threats: if you do not send me the beads, you are not like a father to me/ if I do not like the beads, I will send them back and your efforts will have been wasted. Given the relationship between writer and reader, son to father, this technique demonstrates a clear lack of audience awareness on the part of the writer.

Conclusion

What can we learn from these ancient instructions? First, their existence demonstrates that humans have been writing instructions as long as they've been writing. Why was instructional writing first? Because technology requires instruction, and the explosion of technology in ancient Mesopotamia—agriculture, commerce, accounting, writing, science, and medicine—had to be explained to the user of that technology. Secondly, the linguistic techniques the ancient Mesopotamians developed for giving instructions are very similar to the techniques we still use and teach today. Why are we still using imperative voice, steps, and appropriate levels of detail thousands of years after the Mesopotamians did the same thing? Because these techniques work. Imperative voice removes ambiguity from a task. Logical order makes it possible for someone to duplicate a result, as does the inclusion of specific detail. However, these documents also suggest that ancient writers did more than imitate the structure of accepted genres of writing. The writers I have included in my analysis largely follow the conventions of the genre of instructional writing, yet they also intentionally alter those conventions for rhetorical effect—both good and bad. They practiced rhetoric long before the Greeks, and given the existence of school tablets, it seems reasonable to assume that this practice, at least to some degree, was consciously taught. We know, for example, that students were expected to copy exemplary texts chosen by their teachers. This practice of imitation is further evidence that the Mesopotamians were at least beginning to understand the principles of persuasion. The Mesopotamians understood the need to choose persuasion over convention thousands of years *before* the writing of the instruction manuals Brockman analyzed. Unfortunately, because educational records from ancient Mesopotamia reveal little of the theory behind what writers were taught, we can only surmise what led ancient writers to understand the need for persuasion. We have thousands of student exercise tablets, but we have no textbooks. We have thousands of hand copies of letters that were copied by generations of students, clear evidence that the letters were considered exemplary, but we have no written explanation of why they were exemplary. Perhaps, persuasion is simply an inherent component of human communication. However, while we may not know how the instructional genre was conceived of or taught, or how ancient writers seemed to inherently understand how to make rhetorical choices, we do know these things happened. And we can question the effect this has had on today's practices. Who knows what is waiting under the sand to be discovered?

Chapter 3

Finding Our Missing Pieces

Women Technical Writers in Ancient Mesopotamia

> Women are not a "special issue," but form half the population.
>
> —Halton and Svard, *Women's Writing of Ancient Mesopotamia*

Because your father was a teacher, you are an educated woman. You can write Cuneiform, and you understand the specialized languages for many trades. You have used your skills to start your own weaving business, which you run from your father's house. Now your father is dead, and your brothers want to use the house for their own purposes. They have arranged a marriage for you, so they can place your care in someone else's hands. Your prospective husband does not want a wife who runs a business. Because you can read and write, you remind your brothers that according to the law, if you choose not to marry, your bridal portion can instead be used to purchase your place in a Naditu community. Because you can read and write, a year later you are living in the Naditu community, running your business, and training two of your nieces so that one day they can also be independent if they choose.

The history of women as technical writers is a complex puzzle with many missing pieces. In 1997, Katherine Durack argued that "women are largely absent from our recorded disciplinary past" (36)—an opinion supported by Isabelle Thompson and Elizabeth Overman Smith's 2006 study. Thompson and Smith used a series of key words to identify the number of articles published about women, gender, or feminism between the years 1998 and 2004 in key technical communication journals: *Journal of Business and Technical Communication, Technical Communication Quarterly, Technical Communication, IEEE Transactions on Professional Communication,* and *Journal of Technical Writing and Communication.* While Thompson and Smith (2006) identified eleven articles about women, only four related to the history of women in technical writing. In addition, each of these four articles looks specifically at texts written in English, and only Elizabeth Tebeaux (1998), who analyzed the work of women technical writers from 1641 to 1700, looks at works written before the nineteenth century.

Every piece of published scholarship on the participation of women in the development and practice of technical writing provides a new piece to the puzzle; however, until we identify a finite starting point for women's participation, no benchmark can exist upon which to build our theories. Durack (1997) argues that to find evidence of women's technical writing we must look outside of traditional male spaces—a position criticized by Zainab Bahranie, who argues that such a position relegates women's history to that which is inherently feminine: childcare and cooking (2011). Charles Halton and Saana Svard (2018) support this position, believing the goal of textual analysis should not be to identify the "essentially female features" of a piece of writing but to understand how the web of gender within the women's culture determined their authorship as women (27). Clearly, we should continue to analyze women's texts on women's issues, but we should not *assume* that women historically wrote only about women's issues because an analysis of the first writings produced by women proves this assumption wrong.

The first systems of writing were developed by four distinct cultures: Mesopotamian and Egyptian (3300–3200 BCE), Chinese (late thirteenth century BCE), and Mesoamerican (300–200 BCE). However, only the Mesopotamian culture produced examples of literature known to have been written by women (Halton and Svard 3–4). Sadly, women's writing was not as common as men's; however, "within an environment in which hardly anyone—male and female alike—could write, more than a few women learned this skill, and produced a sizable number of documents" (Halton and Svard 4). Even more importantly, a large majority of the texts written by the women of ancient Mesopotamia were not about traditional women's issues; they wrote for business purposes and instructional purposes. The first recognized author to sign her name to her works wrote to teach women how to use the power of words to ensure their own power. This fact is completely contrary to what most modern scholars believe. Case in point, Tebeaux (1998) argues that it was during the period of 1641–1700 AD that "the first technical works by women appeared" (Loc. 1606)—a fact this chapter will dispute.

By acknowledging that the first women technical writers began writing in the twenty-fourth century BCE, we create a factual and holistic representation of the history of women as technical writers that is much different from those we currently recognize. The inclusion of the women technical writers of Mesopotamia not only extends our history and women's role in that history by eight thousand years, but it also gives primacy to technical genres rather than rhetorical genres, in effect completely revising the history of writing. An analysis of the writing produced by women clearly proves that the first written documents of historical significance were not produced by Homer. Long before the works of Homer were the works of Enheduanna—the first woman writer and the first nonanonymous author ever identified (Halton and Svard 2018). Enheduanna wrote many of the period's great poems, and in the most well-known of her poems, "Hymn to Inanna," she employs the use of an instructional writing style to create a persuasive and intentional argument about the power of the word and how women can harness that power. Her work calls into question our disciplines' commonly accepted beliefs that persuasive writing began with Homer and was conceptualized largely by men. And while easily the most famous, Enheduanna was not the only woman writing at this time. Women were free to participate equally with men in the practice of literacy. First, though, we must understand how scholars of Cuneiform assign gender to recovered texts.

Assigning Gender to Ancient Mesopotamian Texts

Translators of Cuneiform texts have difficulty identifying the genders of the writers of most recovered texts. As Samuel Meier (1991) explains, the lack of a gender marking in the early Sumerian language meant that all scribes, male and female, were simply *dub-sar*. Therefore, Meier reasons it is impossible to know which scribes were male and which female. However, ". . . female scribes are certainly a part of Mesopotamian culture . . . from the 3rd to the 1st millennium" (1991, 541). In fact, textual evidence shows that the education of women as scribes (the technical writers of their day) was not insignificant (Meier 1991). So, while the gender, or even the name of the scribe who wrote a specific text cannot usually be known conclusively, logic tells us that because we have conclusive evidence of women scribes, some of the surviving texts must have been written by women.

According to Meier (1991), males dominated in the communication network, but women were not locked out. One type of textual evidence for women scribes exists in the period's literature. The first patron of scribes, who oversaw their craft, was the goddess Nidaba, also called the queen of scribes:

> On the day, the shepherd-man went out to the plain.
> Where, oh Dumuzi (are you going)?—I will go to the stall.
> His sister, Queen of scribes,
> Was standing (?) there (?) in the open air (?) (Alster 1975, 218, 23–25)

However, despite this clear evidence of women as scribes in the mythology of the culture, evidence of actual woman scribes has been hard to locate. The first verified woman scribe was Enheduanna, daughter of Sargon of Akkad, thought to be the author of a Sumerian poem "In Praise of Inanna" (Pearce 1995, 2266). Although known her for her poetry, Enheduanna was also a priestess. As a priestess, she would have conducted various rituals, a role requiring the use of technical documents containing the words of the rituals and instructions for performing those ceremonies. Another woman, Nishatapada identified herself as a trained scribe in her letter to the king of Larsa (Tetlow 2004) and her elegant prose "became the object of study in subsequent generations of apprentice scribes" (Pearce 1995, 2266).

Rivkah Harris identified evidence of another eight naditu women[1] who worked as scribes during the reign of Hammurabi, including Inanna-amamu, who was a naditu *and* a scribe. Her father Abba-tabum was also a scribe (1962, 8). Another tablet containing the names of thirteen women scribes was found in what is thought to be the naditu cloister of the city of Sippar (Harris 1975). Meier (1991) argues that the number of women scribes proved to have existed suggests that the "educational investment in women was not on a small scale" (542). It is also likely that naditu scribes taught other naditu women in their homes within the naditu community following the tradition of the *edubba*—the scribe school (George 2005). According to Harris (1964), the only administrative position within the naditu community that was a lifetime sinecure was that of scribe, and "there was only one at any given time" (132). One final piece of textual evidence points to naditu scribes training other scribes. A study of contracts from Sippar found a recurrence of the phrase "At the opening of the lattice" (132). Some scholars took this as evidence that the naditus of Sippar lived a cloistered life. However, later studies suggest "its use was probably only a scribal fad without any real significance" (130). For a scholarly vogue to have developed among a specific group of naditu scribes, it had to have developed as an offshoot of their scribal training. Although women had less access to education than men, they did learn to write and work as scribes. And even those women who might not have been able to write effectively used scribes to create the documents they needed to conduct their businesses.

Naditu communities were not the only female communities that included at least one scribe as standard practice. A king's primary consort, or first wife, was the head of her own household. Among the many personnel that served the consort were scribes (Melville 2004). The presence of a scribe would have been necessary because the consort "could own estates (land and villages) outside the palace, and she might even own a palace of her own" (48). The consort would have used the writing produced by her scribe to manage her investments—a great accomplishment dependent upon the use of written documents. Specifically, two classes of women wrote.

Two Classes of Women Writers

Although the women of Sumer (3000–2000 BCE), Babylonia and Assyria (2000–331 BCE),[2] lived in a patriarchal society, they were not without power and influence in both the private and public sector, and women were free to buy and sell whether they were under patriarchal control or not, and they could work as priestesses, all of which required the use of writing.

WOMEN UNDER PATRIARCHAL RULE

Any woman who was married or an unmarried dependent of her father or brother(s) was under patriarchal control. According to Elisabeth Tetlow (2004), even women under patriarchal control had fairly strong financial protection and could conduct business with a third party with the support of a husband or son. Women obtained the money necessary to own a business or buy and sell property in several ways.

When a woman married, her family paid her husband a dowry. If a woman was widowed, she kept her dowry and any personal gifts from her husband. And she was free to use his house, which belonged to their children, as long as she lived. If her husband left her no gifts, she received the same amount as his other heirs. If a woman's father provided her with neither husband nor dowry, her brothers were required to provide for her.

Royal women were also under patriarchal rule—either that of the king, or his sons (Parker 1961, 37). However, like other married women, the king's consorts could conduct business and own property. An administrative tablet from the Northwest Palace at Nimrud identifies a queen, Hasdai, as a *rab alani*, or owner of estates (Parker 1961). Other free women who did not marry became priestesses, which freed their families from the obligation of supporting them and freed them from patriarchal rule.

Women Not Under Patriarchal Rule

In this culture, no woman who was over age fourteen would have remained in her father's house. If the family had the financial ability to pay her dowry, such a woman, whether she was unmarriageable or unwilling to marry, could become a priestess. If her family could not afford the dowry, she could become a profane priestess, or hetaerae (prostitute). Both temple priestesses and profane priestesses were not under patriarchal rule and could conduct business independently.

Temple Priestesses. Temple priestesses either received a dowry from their fathers or inherited part of their fathers' estates—but not both. In the case of a dowry, if the father gave his daughter a sealed document stating that it was hers to do with as she chose, her brothers could not touch it. But without the document, she could not sell any inherited property because it belonged to her brothers. Although all priestesses were free of patriarchal control, different levels did exist among the priestesses:

- En—highest rank. Equal to male rank of en, and second only to the royal rank.
- Sum—second rank. Served one of several gods.
- Nin-dingir—position differed from city to city. (Dianakoff 1986)

Naditum. Naditum were women priestesses who lived in cloistered communities in two major cities, Sippar, and Nippur, during the Old Babylonian period. Naditu of Sippar served the god Samas, while naditu of Nippur served the goddess Ninurta. Most of what is known about the naditu women comes from the Old Babylonia Sippar texts. In nearly 70 percent of the texts, one or more participants is a woman. In 97 percent of the cases where a record exists between a man and a naditu, the naditu was the text keeper (literally owner of the tablet recording the transaction), meaning that she was the recipient of property, the lessor of property, etc. (Stone 1982). However, the numbers must be considered in context because archaeologists recovered the tablets from the cloisters where the naditu women lived, which explains the high percentages (Stone 1982).

Profane Priestesses—Hetaerae. At this time, "prostitute" had a specific cultural meaning. In many ways, Mesopotamian society considered hetaerae profane priestesses unlike temple priestesses who were religious priestesses. Like temple priestesses, profane priestesses were under the protection of the goddess Istar-Inanna. According to Dianakoff (1986), evidence suggests they also had specific rituals and prayers that protected them in their trade as prostitutes. These women probably became profane rather than temple priestesses because they had no one to pay their dowry. Dianakoff posits that many profane priestesses came from homes with no men (231). Even these women used writing for business purposes. We have a textual record of a woman named Baburisat, who was likely a prostitute, selling a baby to another couple (Dianakoff 1986). To legalize the sale, the woman had to use a scribe to write the required document, a copy of which she would have kept. Based on common business practice, we can assume that she would have used her personal seal to "sign" the document of sale. Given the evidence, it seems fair to assume that women of all classes both wrote and used technical writing to accomplish tasks. However, the act of writing was more complex than it is now.

What It Meant to Say a Woman "Wrote"

Modern scholars recognize that writing is a social construct and as such, the production of texts is a cultural phenomenon. However, Martti Nissinen reminds us we must view writing as a cultural *reaction* to the gender system (Nissinen 2013). For example, the gender system in ancient Mesopotamia was patriarchal. However, in this culture even women under patriarchal control had strong financial protection and could independently conduct business with a third party with the support of a husband or son. To conduct business a woman had to write, so the gender system was partially the catalyst for women writing for business purposes. However, the Mesopotamians did not exclusively define "writing" as the act of pressing a stylus into clay to make Cuneiform symbols because a "writer" could be the person who dictated words to a scribe, who then translated those sounds into symbols and placed them on a clay tablet. Dominique Charpin (2010) argues from textual evidence that in addition to scribes, kings and queens, the clergy, members of the military, and both male and female merchants had differing levels of literacy, all of which enabled them to use writing for technical purposes:

- Ability to read and write—possessed by trained scribes
- Ability to read and write at a lower level—common for those with administrative positions
- Ability to read—the majority of the population, who would have used a scribe to "write."

Scribes were certainly the most literate members of Mesopotamian society, but virtually every member of society participated in the practice of literacy to some extent. And we should not make this assumption about men only—women were also literate. So, while we must broaden our understanding of who practiced literacy and how, we do not have to look outside the realm of the public to find women working as technical writers and using technical writing to actively participate in every aspect of their culture. This position is supported by Durack (1997) who argues that writing that accomplishes an action, whether in a public or private sector, is technical "because technical writing exists to accomplish something" (107). However, because the concept of literacy was more complex, we must use new terminology to define those who practiced it.

Instrumental Agent

According to Halton and Svard (2018), a writer is an instrumental agent of the act of writing when she copies existing works. When a scribal student copied a known work of literature, she was an instrumental agent of writing because her copy was instrumental in preserving the work.

A woman who is featured in a text, though she is not the writer, can also be considered an instrumental agent because her role in the text is *instrumental* in the production of knowledge. The recurring role of the goddess Inanna in the period's literature makes her an instrumental agent. Women can also be independent agents.

Independent Agent

A writer is an independent agent of the act of writing when she creates new texts or literary traditions. When a woman wrote, or more often, dictated a letter containing business instructions, or requesting payment of a debt, she was an independent agent of writing. Enheduanna, who wrote many poems, is perhaps the most well-known independent writer of ancient Mesopotamia. These terms provide us with a method for examining texts known to have been written by women, known to have been dictated by women, or known to feature women—a necessary distinction based on the culture's use and understanding of what it meant to write. In addition to identifying the agencies of writing, we must also identify the genres of writing.

Genres of Writing

Documents written by Mesopotamian authors belong in one of two genres: literary or nonliterary (Halton and Svard 2018). Literary texts contain specialized knowledge or material of cultural or religious significance. Literary writing was meant to be eternally preserved and passed down through history. Nonliterary texts included administrative and financial documents and letters. Nonliterary writing was strictly ephemeral and served a purpose only during the life of the transaction that it recorded.

It is important to avoid our modern tradition of assuming the literary genre would not include any examples of technical writing. Halton and Svard believe that we need to give more study to how much the meaning of a text "lies with authorial intent and how much we can construct the meaning of the text based on the actual *use* of the text" (2018, 91). Working from this paradigm of reader response rather than authorial intent, we can use Durack's (1997) categories of use to demonstrate that both literary and nonliterary texts can be considered technical when they allow the agent, whether instrumental or independent, to accomplish something, provide instruction, or make the tacit explicit. I chose each of the documents analyzed in this chapter because they meet one of these criteria:

- Accomplishes something—women scribes wrote letters, instructions, business receipts, and legal documents. These documents were essential to women's accomplishments as entrepreneurs, landowners, and priestesses.

- Provides instructions—women, who have been identified as doctors and midwives, wrote and used instructions for treating the illnesses of children, women and the elderly. They also appear as the *sekrum*, or wise woman, a literary figure who gives instruction.

- Makes the tacit explicit—goddesses provide explanations for the origins of life, and quest for immortality. And grave goods found with Queen Puabi provide evidence that the written word could protect the soul.

In the analysis that follows, I identify each document as the work of either an instrumental or independent writer, and as either literary or nonliterary. I base its definition as a technical document on whether it meets one of Durack's categories of use.

Literary Texts Written by Independent Agents

Although the vast majority of recovered Mesopotamian texts are examples of routine business transactions, and as such, have no authorship, of the small body of literary texts known to exist, the Princess Enheduanna either authored or collected some of the most significant (Halton and Svard 2018, 51). Current histories of writing almost exclusively privilege the works of men: Homer, Pliny, Plato, Aristotle, Cicero, and Quintilian are just a few of the illustrious men who have been credited with writing the first documents of signficance, yet each of these men was preceded by a woman who wrote.

Enheduanna's importance cannot be overstated: she is "not only the first attested female author in world history, but she is the first non-anonymous author known, male or female" (Halton and Svard 51). In what is believed to be a political move, her father, Sargon I, King of Akkad, made her the high priestess of Ur. In this position, she wielded great power and would have led rituals, administered the vast temple complex, and functioned as a diplomat on behalf of her father (Halton and Svard 51). In one of her greatest works, "A Hymn to Inanna," Enheduanna not only produces one of the first known works of literary merit, she uses her poem as a persuasive vehicle in which she argues that women are not only equal to men but superior. Perhaps even more significantly for the field of technical writing specifically, she uses instructional writing to make her argument, which means that technical writing, rather than the rhetorical works of the Greeks and Romans, can claim to be the first example of intentional persuasion. Such a conclusion could potentially lead to a complete reconsideration of when and where the practice of intentional persuasion began, a topic I explore in a later chapter.

"A Hymn to Inanna"—To Accomplish Something and Provide Instructions for Doing So

This poem, though literary, can be considered technical because it allowed Enheduanna to accomplish a specific purpose, and it gave her a method of providing women with instructions for how to do the same. In her poem, "A Hymn to Inanna," Enheduanna does more than just praise Inanna's power; she specifically argues that Inanna is the greatest of all the deities, surpassing even the male deity An, normally recognized as having the greatest power in the Sumerian pantheon:

> Magnificent queen who gathers the divine presence of heaven and earth
> And rivals the great An,
> The greatest among the gods, she validates their verdicts
> The Anuna gods crawl at her magnificent word,
> Its path An does not know, he does not go against her command.
> She changes her plan, and no one knows it. (Halton and Svard 80)

In this portion of the poem Enheduanna uses powerful verbs to describe Inanna's power—a method still used in modern instructional writing. She *validates* the other god's verdicts, suggesting that without her validation, their verdicts would not stand, which establishes her as the ultimate power. The other gods *crawl* at her command. An cannot *follow* the path of her logic, and so cannot *go against* her words of command. Because they cannot understand her words, the other gods cannot anticipate her actions and cannot know her mind. Clearly, Inanna's power is directly linked to her "*magnificent word.*"

Although later in the poem, Enheduanna does praise Inanna's physical prowess, the fact that she begins her poem by praising the power of Inanna's *word* makes it clear that her ability to use language is most responsible for her ability to supersede the other gods. While Enheduanna's intention in writing this poem is unknown, it seems naïve to assume that she did not intend to proclaim the power of a woman and to call to other women to use their words to accomplish great deeds of their own by emulating her actions, which meets Durack's (1997) criteria for "accomplishing something." Later in the poem, she more specifically describes the power of Inanna's word and lists the calamities that will befall the men who do not heed her voice, in effect, providing women with instructions for how to use their words to wield power:

> At her imposing voice the gods of the land become scared.
> Her roaring makes the Anuna gods tremble like a single reed.
> At the sound of her everyone hides together.
> Without Inanna the great An cannot make a decision and Enlil cannot
> Determine fate.
> The mistress who raises her head over the mountains and is supreme, who
> Attacks her? . . . .
> Fear of her gives people tremors and causes them to flush like a man
> Possessed by a demon.
> She brings confusion to those who are disobedient. (Halton and Svard 2018, 80)

Enheduanna again emphasizes Inanna's *imposing voice*, which *roars*. Not only does her voice create fear in others, but it also empowers two male gods, An and Enlil, to complete their duties as gods. Does she instruct them? Even more intriguing, Enheduanna describes Inanna as "mistress," the title used to describe the woman of the house. I believe this was an intentional word choice meant to empower women by establishing them as the true rulers of the home who have the power to "bring confusion" to members of their family who disobey them. Enheduanna, a female priestess in a patriarchal society, used her power as a writer to loudly proclaim that it is women, not men, who truly rule the world. It is the woman of the family who has the power to validate her husband and son's decisions. It is the voice of the mistress of the household upon whom the men depend when making decisions or determining the future of their family.

Based on Durack's categories, we can clearly argue that this poem is technical. This poem, which was written by a woman, copied by hundreds of scribal students, and read by thousands of Sumerians, certainly achieved the goal of establishing women (even within this patriarchal society) as a source of power in their own right. This poem is evidence that women did write about topics other than those related to women's issues. As this poem establishes, Enheduanna wrote about the power of the word in order to empower other women to find their own powerful words.

Instrumental Literary Texts

Mesopotamian culture had a rich mythology in which their gods and goddesses carried out great deeds. The purpose of these stories is to explain the human condition in explicit terms.

The Atra-Hasis—To Make the Tacit Explicit

Written around 1646 to 1626 BCE, the *Atra Hasis* tells the story of the creation of humans and is one of the great works of Mesopotamian literature. It was widely copied in scribal schools until well into the first millennium BCE. This story attempts to explain human origin. In the story, the goddess Mami, after creating humans, issues instructions regarding their behavior:

> In the house of a woman who is giving birth the mud brick shall be put down for seven days. Belet-ili, wise Mami shall be honored. The midwife shall rejoice in the house of the woman who gives birth and when the woman gives birth to the baby, the mother of the baby shall sever the cord herself. A man shall cleave unto a woman, a boy to a girl. A girl shall be ready by

> the sign of her bosoms, a young man by the beard upon his cheek. In the gardens and the waysides they shall cleave unto each other, a wife and her husband shall choose each other. (Lambert 1991–1993, 69)

Significantly, it is a woman, Mami, who creates humans, and at their birth, the mother severs the cord herself. And Mami specifies that a man shall cleave to a woman, a boy to a girl. It is the man who is to cleave, adhere faithfully and loyally to his wife. Mami literally places the action of subservience, cleaving, on the man and not the woman.

It is also clear that women, not just men, had the right to choose a mate because "they will choose each other." Mesopotamian society was patriarchal, but this work, considered *one* of the great (if not *the* greatest) works of Mesopotamian literature, clearly suggests that women had, and more importantly were encouraged to have, a strong element of autonomy in their relationships.

The Old Man and the Young Girl—to Provide Instruction

Two texts from the Old Babylonian period, *The Old Man and the Young Girl*, and *The Three Ox-Drivers from Adab*, feature as one of the primary protagonists, the *sekrum*. According to Alhena Gadotti (2014), the Mesopotamian literary canon includes the role of the *sekrum* as a literary tradition. Some scholars have translated *sekrum* as "cloistered woman" (Alster 2005, 385–86), while others as "court lady" (Gadotti 2014, 66). Whether *sekrum* means naditu or court lady, one point is clear, the *sekrum* is a "wise" woman. In both texts, a king asks the *sekrum* for help in solving a problem. Sadly, these texts are badly damaged, but enough of *The Old Man and the Young Girl* exists to verify the role played by the *sekrum*—to offer a solution to a problem the king is unable to solve:

> The king did not answer,
> (But) entered (the abode of a) court lady [*sekrum*].
> And repeated the speech (of the old man) to the court lady.
> "My blood was fine, my . . . , . . . , (and) . . .
> "were multicolored. My . . . barley had no god." Thus he said.
> The court lady replied to the king:
> "My [lord], suppose the old man took a young girl as a wife.
> [. . .] and the old man will regain his youthfulness. (66)

In this example, the wise woman suggests that the man can solve his problem by marrying a young woman. The problem faced by Gilgamesh is more complex.

The Epic of Gilgamesh—to Make the Tacit Explicit

Chapter 4 of *The Epic of Gilgamesh*, "The Search for Eternal Life," includes another example of the *sekrum* as advice giver. Overcome with fear of death, Gilgamesh seeks immortality. On his journey, he meets the young woman Siduri [the *sekrum*]:

> She answered, 'Gilgamesh, where are you hurrying to? You will never find that life for which you are looking. When the gods created man they allotted to him death, but life they retained in their own keeping. As for you, Gilgamesh, fill your belly with good things; day and night, night and day, dance and be merry, feast and rejoice. Let your clothes be fresh, bathe yourself in water, cherish the little child that holds your hand, and make your wife happy in your embrace; for this too is the lot of man.' (66)

In this passage, the *sekrum* addresses the human condition, and encourages Gilgamesh, who fears death, to cease his search for eternal life, which belongs only to the gods, and to embrace the life he has as a human. His search for immortality is made physically explicit as he follows the directions provided by the *sekrum*: ". . . down in the woods you will find Urshanabi, the ferryman of Utnapishtim. With him are the holy things, the things of stone. He is fashioning the serpent prow of the boat. Look at him well, and if it is possible, perhaps you will cross the waters with him" (66). Ultimately, Gilgamesh's quest fails, as the *sekrum* had foretold, and he returns to the city of Uruk to share his story of man's quest for immortality and its failure with others. Clearly, this story was meant to encourage readers to accept their lot as mortals and embrace life. Humans at this time required more than answers to philosophical questions; they also needed help with physical problems.

Many incantations, used for both personal and community purposes, were written in the Old Babylonian period. The importance of these texts lies in the fact that many of them are found in multiple collections. In fact, by the first millennium, "The great series had been canonized"

(Goff 1956, 5), which again proves that the first literary canon was not written by Homer. It is because they were intentionally preserved that these incantations are considered literary rather than transactional, though their purpose was clearly practical in nature. It is in instances like these that we must remember Halton and Svard's recommendation to find a text's meaning not in authorial intent but based on its use (2018, 31). These incantations provided instructions for healing both individuals and communities.

Ritual to Induce Labor—to Provide Instruction

In the Ancient Near East, rituals served many purposes, including medical treatments. Again, we cannot verify who wrote the rituals presented here, but evidence exists that shows women did work as *asu* (doctors). In the Ur III period, a doctor has a female name. The names of two women doctors were found at the court of Mari, and an Old Babylonian list references a woman doctor and a woman midwife (Stol 2016, 371). We also have textual evidence of wet nurses. The inscription on a seal identifies the owner as Daguna, the wet nurse, daughter of Tisa-Dimmuzi, the attendant (Collon 1987, 21). Given this information, it is not a stretch to assume a midwife could have written and certainly used instructions such as these to induce labor: "Recitation: Run to me like a gazelle; flee to me like a little snake. I, Asalluhi, midwife; I will receive you" (Scurlock 2014, 129). The presence of the midwife's name in the recitation also suggests that the seal was owned and used by the midwife, Asalluhi, herself. Just as rituals to start labor were necessary, so were rituals to stop it.

Ritual to Save a Pregnancy—to Provide Instruction

Instructions for rituals, such as this one, were meant help a pregnant woman keep a child:

> Stones of white KA you will string on a band of white wool, *pusikku*,
> Five fingers between them you will attach; a cylinder seal of *haltu* stone,
> A cylinder seal of *subu* stone, *su* stones male and female at her neck you will place;

> A *subu* stone of the right to her right hand you will attach;
> a *subu* stone of
> The left to her hand you will attach;
> A "date stone," a SAB stone, a *musu* stone, an *asgigi* stone to
> her waist you
> Will attach; a *nibu* stone, a *hulalu* stone to her right foot,
> A carnelian stone, a stone of lapis-lazuli, to her left foot, these
> two stones
> On a cord
> Of blue wool you will string; in wrappings of red wool you
> will surround
> Them; amid the stones of the hands and feet
> A . . . . you will make. (Such is) the binding (to preserve) a
> pregnant
> Woman from losing her fruit. Example of . . . (Goff 1956, 18)

The majority of the rituals recorded were not just for women.

RITUAL FOR TREATING RHEUMATISM—TO PROVIDE INSTRUCTION

Instructions for treating rheumatism were also common. Again, though we cannot prove that a woman followed these instructions, it seems safe to assume that women of the naditu cloister would have had rheumatism and been treated by a priestess but also that such a domestic problem might have been treated by female doctors, who have been proven to exist:

> When the tendons of his leg suddenly consume him, he cannot stand and walk, (then this is) a Sa, Gal sickness of two years.
> Draw a Lower World river of bitter corn meal, [in] the river put a Sab, Sur reed fill a Gis-Bar (?) with corn,
> Set (it?) on the Sab.Sur reed, cause the patient to set himself upon it, fill a measure with bitter corn,
> Set (it?) on the Sab.Sur reed, lay his sick foot upon it, rub his foot with bittern corn dough
> [Incantation formula not understandable.]

> The cow with her horn, the mother sheep with her wool, the river of the Lower World with its bank. Command! NN son of NN, may become well! End of incantation.
> Incantation against Sa. Gal.
> Ritual for it: This incantation shall you recite along with the wiping of the leg, that dough
> Lay in a hole in the West, with clay which is mixed with straw its opening close, with a seal of subu stone and Kur. nu [dib] stone
> Its opening seal, his leg purify with a torch. Raise him up, grasp his hand,
> The river of the Lower World, which you have drawn, let him step over seven times and seven times. When he steps across it shall you thus say: [missing]
> In a hole in the West lock it up, and then
> With seals of *shubu* and *kurnu* stone you shall seal the opening. (Goff 1956, 28–29)

As the instructions make clear, a hole was dug, a substance was placed in that hole, and then it was sealed with two specific types of cylinder seal. Interestingly, the practice of sealing a business transaction is the same used for this healing ritual—clear evidence that the Mesopotamians placed huge significance on the act of witnessing any act of communication, whether transactional or ritualistic.

Kislimu Ritual for Esagil—to Provide Instruction

A set of ritual instructions in a late Babylonian Kislimu Ritual for Esagil requiring a naditu priestess to write, provides proof that women did write. The text comprises instructions for conducting the ritual for days three and four. Several lines of this document specifically state that a naditu priestess must write:

> 101 A naditu-priestess . . . her reins hitched up,
> the curtains drawn,
> 102 And seated . . . she will put the alu-drum
> by the barley beer, she will write seven
> inscriptions. (Cagirgan 1991–1993, 98)

These two lines are clearly technical in nature as they provide instructions for a naditu priestess who was required to conduct the Ritual for Esagil as one of her duties, and they provide further evidence that women wrote. And women didn't just write; a woman was the goddess of writing.

Despite Plato's contention that we thank the Egyptian God, Theuth, for inventing "number and calculation, geometry and astronomy. . . . and above all writing" (Plato 1956, 68), we should not credit a god at all, but a goddess. Both artifacts and textual evidence from ancient Mesopotamia—3200 BCE to 75 BCE (Charpin 2010, 7)—offer demonstrable proof that a Sumerian goddess was first linked to the act of writing, and it is she who can claim the title of first technical writer.

Praise Be to Nidaba—To Make the Tacit Explicit

The goddess Nidaba was first the goddess of grain and then of writing. Consequently, her primary role as a deity was as the patron of scribes. In the Sumerian period (3000–2300 BCE), literary texts often ended with the prayer, "Praise be to Nidaba" (Tudeau 2016). Although we have less textual evidence of Nidaba's writing than other goddesses such as Inanna, in the epic poem "The Cursing of Agade," she is included in a list of "the great gods" (2001). Most significantly for the current study, Nidaba, as a goddess, personifies one of the most "fundamental developments in the creation of Mesopotamian culture" (Tudeau 2016). As first the goddess of grain, a deity of great importance to the people of an agrarian society, her metamorphosis from goddess of grain to goddess of writing is indicative of the growing importance of record keeping—accounting—as the agrarian society became an urban society built on commerce (Tudeau 2016). While the fertility of their crops, including grain, was of primary importance as people learned to grow crops to support a sedentary rather than nomadic lifestyle, once the growing of crops was established, and surplus crops were being consistently produced, the need for recording this new wealth became of equal, if not greater, import. So, the successful cultivation of grain was the catalyst for the development of the first genre of writing, and that genre was technical—not literary. Thus, Nidaba is not just the goddess of writing—she is the goddess of technical writing and the patron of the members of an elite profession—technical writing. Her importance as a "guardian deity is demonstrated in this child's school composition:

"Young man, because you did not *neglect* my word, did not forsake it,
May you reach the pinnacle of the scribal art, achieve it completely.
Because you gave me that which you were by no means obliged (to give),
You present me with a gift over and above my earnings, have shown me great
Honor,
May Nidaba, the queen of the guardian deities, be your guardian deity,
May she show favor to your *fashioned reed*,
May she take all evil *from* your *hand copies*. . . .
Nidaba, the queen of the place of learning, you have exalted."
O Nidaba, praise! (Kramer 1949, 206)

Nonliterary Texts Written by Independent Agents

Women scribes wrote letters, instructions, business receipts, and legal documents. These documents were essential to women's accomplishments as entrepreneurs, landowners, and moneylenders. The existence of these documents also gives the lie to any assumptions that women were historically barred from engaging equally with men in business and commerce. Their existence, which seems to have been an accident of the strength of the material on which they were written, begs the question of whether other such documents might have existed in later periods, such as that of ancient Greece and Rome or Medieval England, and been lost. Perhaps women were writing letters for business. Perhaps women did own receipts that legitimized their ownership of property. Perhaps our assessment of how actively women participated in the transactional writing of their cultures has been colored by our unwillingness to credit anonymous documents to women. Perhaps the documents don't exist because the paper on which they were written was destroyed or repurposed. Receipts do not have literary value; consequently, they are not preserved beyond the life of the transaction. Clearly women in ancient Mesopotamia did write. Why isn't this same level of literary activity among women present in later periods? Or was it?

LETTER WRITTEN BY NINSATAPADA TO RIM-SIN,
THE KING OF LARSA—TO ACCOMPLISH SOMETHING

Ninsatapada was the daughter of the ruler of Uruk and Durum and a high priestess in Durum. After Rim-Sin conquered her cities, in the process causing her to lose her place as priestess, she writes to ask him to return her position to her. In this highly eloquent letter, she relies on a combination of flattery and logic to make her argument—further evidence that the teaching and use of rhetorical strategies predates not only Homer but also the Greeks and Romans. Because we know that Ninsatapada was an educated scribe and wrote this letter herself, we must assume that the persuasive techniques that she employs were part of what she was taught in the *Edubba* (scribal school):

> Say to my king:
> Rim-Sin, the young supervisor who works for Enlil, the faithful shepherd, faithfully given to save the whole nation, all-knowing, exceedingly wise, assimilating everything, the advisor of dispassionate wisdom whose value none can fully see, the true judge who loves the righteous person like Utu himself. (Tetlow 2004, 24)

Although this letter predates Aristotle by thousands of years, she anticipated his advice to "render the audience receptive" by effectively praising the king and his ability and wisdom as a true judge and righteous person—the assumption being that such a person will not fail to see the logic of the argument she is about to present:

35–42

> Look favorably upon me, let your pronouncement brighten this dark day. They have made me live like a slave these five years away from my city. I have nothing. Because of your silence, my countenance has changed. My body is dead; my course is bent. In the deserted place I clap my hands, I do not know . . . Though I am youthful in old age, I am abandoned, I am driven from my bedroom. Like a bird in a cage with its young gone from its nest, my children are scattered afar. I do

not have anyone to work for me. They do not clammer for my home—they moan about it like doves.

43–47

My food fills me with lament. I cannot be calmed. Life is interminable. They gossip about my calamity. I have become an insulted woman. As to my position, they have made me a servant. Give ear to this situation! My female servant will not make me a garment while I am dressed in rags! Who will speak for me? (Halton and Svard 2018, 101–2)

In the body of her letter, Ninsatapada argues that her status has been taken from her, reducing her to a slave. Having clearly argued that Rim-Sin's taking of her city was justified by his military might, she asks that he use that same power to re-instate her to her rightful place. This letter is a clear example of an elegant, logical, and persuasive argument, the product of an educated woman.

Letter from Serua-eterat to Assur-Sarrat—
to Accomplish Something

The daughter of the king of Assyria was another lady of letters. She wrote prolifically and took an active interest in the education of the women in her family, arguing that the inability to write brought shame to a girl's family:

Why do you not write me any letters, why do you not send me any oral message? Isn't it in reality because people might say: "Perhaps that one (i.e., the writer of this letter) is higher in rank than she." After all: I, Serua-eterat, am the eldest daughter born in the official residence to Assur-etel-ilani-mukinni (= Esarshaddon), the great and legitimate king, king of the world, king of Assyria, while you are only a daughter-in-law, the lady of the house of Assurbanipal, the eldest son of the king born in the official residence of Esarhaddeon, King of Assyria. (Oppenheim 1967, 158)

This letter, from the archives at Nineveh, is perhaps more telling than any of the other texts considered. How can we question the fact that women wrote, wrote often, and were trained to write correctly, when we have a letter such as this? Not only does the writer reference multiple letters, she offers a logical explanation for why she believes the lady Assur-Sarrat has not responded to her. While there may be some validity to her claim that her higher status is the cause, her clear attempt to justify the absence of letters—and, we assume, assauge her own hurt feelings—is representative of human emotions, past and present. We can only assume that her purpose in writing this letter was to accomplish one specific task: shame the receiver of the letter into responding to her queries. The next letter, also written by Serua-eterat, provides a possible answer to Assur-Sarrat's failure to write.

LETTER FROM SERUA-ETERAT TO LIBBALI-SARRAT—
TO ACCOMPLISH SOMETHING

In this example, Serua-eterat chastises Libbali-Sarrat for her failure to do her "composition home work," which may explain the failure to send letters referenced in the previous letter. In the context of the letter, the importance of learning to write on a clay tablet suggests Serua-eterat may have done her own scribal work:

Obv. Lines 1–2

Word of the king's daughter to Libbali-sarrat.

Obv. Lines 3–6, rev. lines 1–7

Why do you not write on your clay tablet? (Why) do you not rehearse your exercise tablet? Otherwise, they will say: "Is this the sister of Serua-eterat, the eldest daughter of the Succession Palace of Assur-etel-ilani-mukinni, the great king, strong king, king of the universe, king of assyria?" And you are the daughter, the daughter-in-law—the Lady of the House of Assurbanipal, the great crown prince of the Succession Palace of Esarhaddon, king of Assyria. (Halton and Svard 2018, 150)

In the letter, she again attempts to shame her sister-in-law for failing to write. She establishes her own importance, "Word of the king's daughter" (150) and then argues that others will judge her (and by association her family) harshly if her sister-in-law does not meet the expectations of her position and learn to write. Serua-eterat shows a clear understanding of the rhetorical power of repetition to accomplish a goal, using no less than four titles to describe the importance of her sister-law's position: and thus the weight of her responsibility to write on her clay tablet!

Nonliterary Texts Written by Instrumental Agents

According to Martin Stol (2016), "In the Old Akkadian period, From the queen down to ordinary citizenry, women, whether married or unmarried, Sumerian or Akkadian, were free to participate in public life on a par with men" (383). And because the conducting of business was dependent on the use of writing, women in ancient Mesopotamia worked with scribes to craft business letters, contracts, and receipts, which they then sealed with their personal cylinder seals, as in this example from a tablet recovered from Tell Rimah. This is an example of instrumental agency because the scribe who transcribed the message signed it. As the text states, the "lady" who received the letter sealed it: "The tablet on the state of the weavers when I established their accounts, I gave to my lady. May my lady seal it now with her seal and she have it brought to me with the weavers."(Charpin 2010, 35). This letter clearly proves that the woman receiving the letter had the authority to verify the accuracy or legality of business accounts by "sealing" the tablet.

Terms of Redress—to Accomplish Something

According to Charpin (2010), it was common for a Babylonian king to order an economic redress (*misarum*). According to the terms of a redress, "Creditors were obliged to take their contracts before a commission, where the tablets were solemnly broken," (Charpin, 2010, p. 49) as the following text describes:

> On one such occasion, in year 13 of Hammurabi, a creditor declared that she had lost her tablet. So that she could not subsequently enforce that contract against her debtors if she happened to find it again, a clod of dirt (kirbanum) was broken as a substitute for the lost tablet, whose value was rendered void as a result. But at the same time, in order to better

protect the debtors, a text was delivered to them describing the ceremony and specifying that if the creditor subsequently found and attempted to enforce a tablet, it would be considered counterfeit and would be broken. (49)

As this document makes clear, women did actively and successfully engage in the business of lending money—clearly to many people. Just as clearly, some of these women were tempted to steal from their clients.

Death Ritual of Queen Puabi—Making the Tacit Explicit

For ancient Mesopotamians, one of the most important technical uses of the word was to invoke the name of the dead to ensure that the person's soul wasn't left to wander for eternity. This use of the word to name humans was an extension of humans' understanding of the use of the word to name commodities—a cognitive leap that required humans to understand their own existence in abstraction. According to Denise Schmandt-Besserat (2007), in 3100 BCE, people still wrote primarily for accounting purposes. However, around 3000 BCE accountants in the city of Uruk began something new—the recording of the personal name of those involved in the business transaction. This occurrence was significant for two reasons: first, to transcribe personal names scribes developed phonograms that imitated the sounds of the name—the first step toward a phonetic language; secondly, the names had ideological significance: "From Sumer to Babylon, a name was considered to be the essence of an individual" (Schmandt-Besserat 2007, 1109). "In particular, people believed that things came into existence by naming them" (Schmandt-Besserat 2016, 3). Naming was an essential element of funeral rituals and ensured the deceased was not lost for eternity. Queen Puabi's grave goods demonstrate that women ruled and used the written word to preserve their souls.

The regular utterance of the name of the deceased was necessary to prevent the person from returning to haunt the living (Schmandt-Besserat 2016). Before the Late Babylonian period, most individuals would have relied on the use of a monthly funeral ritual, the *zakir shumi*, at which a person's child would say the name of the deceased the required number of times necessary to "keep the name alive" (Bayliss 1973, 117). However, according to Schmandt-Besserat (2016), the intent of the Ur texts, excavated at the Royal Cemetery of Ur, and dated to the second quarter of the third millennium, was to supplement or replace the all too fallible *zakir shumi* for the utterance of ancestor names. Meskalamdug, and all the Ur nobles buried with extravagant wealth, believed that because the phonetic signs reproduced the sound of their name, writing had the awesome power of a personal utterance. This is particularly credible when one realizes that, at the time, people always read aloud; therefore, the signs always evoked sound to the reader. Furthermore, people believed they could listen to a tablet because tablets had a mouth. The ability of an inscribed object to "speak" in conjunction with the new technology of inscribing names provided new safeguards for both the living and the dead as evidenced by the Ur artifacts.

Among the Ur artifacts was a seal with the inscription "Puabi, Queen" (Schmandt-Besserat 2007, 1185). The Ur inscriptions, unlike most texts, were not inscribed on clay tablets but on vases and bowls of gold, copper, and shell, and seals of lapis lazuli—"hard and durable objects that (except for copper) did not corrode nor tarnish" (1221). These objects clearly represented wealth and prestige, but more importantly, permanence (1221). As Schmandt-Besserat explains, the Ur inscriptions mark a recognition of the awesome power of writing as a means to permanently capture the sounds of speech. "And given the importance of preserving names in Sumer, as illustrated by the ritual of regularly pronouncing aloud those of the deceased, it is logical to assume that casting the ephemeral sounds of names into a permanent form by couching them in writing was conceived as equivalent to a perpetual utterance. The new function of writing was supplementing, or even replacing, the role of the *zakir shumi* (1245).

The inscribing of the name in conjunction with the banquet imagery (see figure 1.10) was intended to act as the monthly banquet, at which prescribed food and drink were consumed as her descendants uttered her name. Puabi, queen in her own right, no longer had to trust her place in eternity to a mortal—the written word protected her, which leads to an interesting conjecture: Was naming one of the first forms of technical communication? At what point in the development of the written word did humans consciously understand the enduring power of the written versus the spoken word?

Conclusion

The special issue on feminism and technical communication published by *Technical Communication Quarterly* in 1994 was only the second attempt to address the absence of women in our discipline's history (Flynn 1997, 313). In the introduction to this special issue, Elizabeth Flynn (1997) argues that the issue demonstrates "the importance of historical investigations of women's contribution to the field, a legitimation of a feminist approach to technical communication, and an exploration of the ways in which focusing on women's contributions changes the field in important ways" (314). The women published in this issue set in motion a movement to study technical communication from a more complete, more inclusive perspective, and while some technical communicators continue this work, much work is left to be done. Although we should continue to analyze women's texts written about women's issues, we should not assume that women historically wrote only about women's issues. A large majority of the texts written by the women of ancient Mesopotamia were not about traditional women's issues; most often they wrote for business: contracts, instructions, and legal documents. This fact is completely contrary to what most modern scholars believe, and it justifies a reconsideration of the many assumptions we have made about the history of women's writing. For example, what occurred between the centuries in which Mesopotamia was the seat of civilization, and the period in which writing was appropriated by the Greeks and Romans, and the Western tradition as a whole? Why did women write so abundantly about the same technical topics as men in the East, and then find themselves barred from writing almost exclusively in the West? These questions deserve answers.

This chapter focuses on a period significantly earlier than any previously studied by the field of technical communication; it attempts to continue broadening not only our understanding of the origins of writing and communicating for technical purposes but also the significant role women played in this development. Women of the ancient Near East were technical communicators. They wrote for business purposes and instructional purposes; they served as writers to the gods and were the gods of writing. Writing developed as a technology because of the need for humans to communicate for professional purposes; it did not develop out of the need for creative expression. And women, not just men, played a fundamental role in the revolution.

Chapter 4

Decentering the History of the Writing Center
A Case for the Mesopotamian *Edubba* as an Early Writing Center

> The creation of a learning culture within a writing center requires "long, future-oriented views of our work and the constant conviction that past, present, and future always inform one another."
>
> —Geller, *The Everyday Writing Center*

You are a young person of eight or nine. You are sitting on the dusty ground of a courtyard. The hot sun is on your shoulders, and a warm breeze carries the scent of ripening barley from the nearby fields. In your hand you hold a piece of wet clay that you have pressed into a round flat shape similar to a lentil. In your other hand, you hold a piece of reed that has been sharpened to a point. You are pressing the angled end of the reed into the wet clay of the tablet trying to mimic the marks left by your teacher. His marks are clear and precise. Yours look like hen tracks. You sigh deeply. Your stomach rumbles, and sweat soaks your back. Will this day ever end?

I have gazed at tablets covered in Cuneiform symbols in many museums. Though fascinating in an abstract sense, those tablets never spoke to me—until recently. I suddenly thought of the many Cuneiform tablets I've seen, and I began to ask myself a question: Who taught students in ancient times how to write using those symbols? Someone must have. Cuneiform is a complex system. Asking that question led me down a rabbit hole and into an exploration of the *edubba*—the Mesopotamian scribal school. The more I learned, the more I realized that these ancient schools, while strange in many ways, also bore a striking resemblance to a place I know well—the modern writing center. But how could that be possible? Such an idea contradicted everything I thought I knew about the history of the writing center.

Didn't the writing classroom come *before* the writing center? Didn't the first writing center develop somewhere in North America? Isn't the writing center the solution to the problem of a one-size-fits-all method of teaching writing? The answer to these questions is no.

The study of the *edubba,* the ancient Mesopotamian school for scribes, is a way to understand one counter-history of writing centers and classrooms—and this counter-history will help to erase the artificial barrier that has been erected between classrooms and writing centers by showing that while the two locations may use different methods, they engage in one practice—the teaching of writing—and are both part of the same singular history.

The history of the teaching of composition has long privileged a narrative in which the Greeks developed the first system for teaching students how to read and write with the goal of preparing them to be working professionals and intellectuals. When writing centers finally came along many thousands of years later, they were thought to serve a singular purpose—the remediation of unprepared students. This job was so distasteful to most centers of higher learning that they tried to look the other way and pretend they didn't exist. In fact, in 1960 Dartmouth College closed their writing clinic, which opened in the 1930s because the school's administrators didn't believe that providing writing remediation was a fitting activity for a distinguished school (Lerner 2009). Sadly, as most writing center professionals understand, this attitude has lingered like a bad smell.

In reaction, Neal Lerner (2009) began a quest to "figure out where the first writing center was created—and how and why" (15). Lerner identified the precursor of the writing center in the laboratory methods that developed in the 1890s, and ultimately identified "one of the first solid references to stand-alone writing centers . . . from a 1928 national survey by Warner Taylor of the University of Wisconsin" (28). Lerner's findings are important pieces of the history of the *American* writing center; however, since writing was invented, writing has been taught. Had I not been drawn out of my own colloquial history by the intervention of a

Cuneiform tablet in a museum, my quest might have ended in a similar location, but my journey has led me down a different path.

The first writing center, the *edubba*, was not what Lerner located at the University of Wisconsin in 1928—a supplement to the writing class. The writing center *was* the writing class. The original source that Lerner was seeking was the *edubba*—the place where activity was *centered* on the teaching and learning of writing. Lerner's quest was limited because he searched within his discipline's own well-established borders.

The time has come to stop privileging a Western-centric knowledge base and begin exploring the "different, complex, but well-established traditions in both writing research and writing construction of other countries" (Donahue 2009, 214). Amy Hodges, Lynne Ronesi, and Amy Zenger all work in writing centers located in the Middle East and North Africa (MENA). While they recognize the value of the knowledge produced in the United States, they also understand that current consultations between the United States and MENA will not be truly useful to either participant until the relationship is "structured as an exchange between equal partners instead of a consultancy that suggests knowledge only flows from West to East" (2019, 50–51). They argue that the solution is a reversal in thinking suggested by Christiane Donahue, who asks that we "begin thinking about where our work fits in with the world rather than where the world's work fits into ours" (2009, 14) because an understanding of "the many traditions of researching writing and the teaching of writing" (Hodges, Ronesi, and Zenger 2019, 45) is necessary to the globalization of our centers, and I would add, our history.

Malea Powell (2002) blames this myopic view on our discipline's continued use of the rhetorical tradition to determine the boundaries of its history: "Typically, the tradition begins with the Greeks, goes Roman, briefly sojourns in Italy, then shows up in England and Scotland, hops the ocean to America and settles in" (397). As Powell's travel itinerary demonstrates, rhetoric as a discipline is "anchored by Western patriarchal values" (Royster and Kirsch 2012, ch. 3). This fixation on an omnipresent tradition can be traced to the work of comparative rhetoricians such as George A. Kennedy, who have determined that all non-Western rhetorics can only be compared to what they consider *the* universal rhetoric—the rhetoric of Western figures such as Socrates, Plato, and Aristotle and their almost exclusively male ancestors. The object of the comparison is to discover commonalities.

Unfortunately, such an approach naturally demotes all non-Western rhetorics—their value lies only in their similarity to Western rhetoric rather than their differences *from*, which perpetuates the belief that knowledge only flows from West to East" (Hodges, Ronesi, and Zenger 2019, 50–51). As Jacqueline Royster and Gesa Kirsch (2012) explain, this binary of Western/non-Western, normative/different has effectively established a criterion for worthiness based on conformity to Western standards (ch. 3). Krista Ratcliffe (2010) describes this problem as "cultural imperialism, which risks relegating the 'excess' of non-Western rhetorics to the status of unimportant at best, and invisible at worst" (205). To address this issue, Royster and Kirsch (2012) offer "globalization" as a methodology. Specifically, they argue that how we connect our historical dots is relevant to the history we create. Connections must be drawn not just "between the classical tradition of Greece and Rome and our own contemporary scene in the United States" (ch. 2) but across the globe, time, and space because a failure to do so will impede American scholars' ability to "hear" work from outside our borders. And this problem affects both theory and praxis.

When Jackie Grutsch McKinney (2013) analyzed the content of recent tutoring guides, she discovered that they share a common assumption: ". . . the tutor and student will likely be white, of high ability, young, and American" (71). Even more significantly, she argues, "If we paint students in this light in our official discourses—tutor training guides—it's likely that we are doing so in unofficial discourse too" (72). How can we tell the complete story of the writing center and its history if we work from the premise that writing center tutors will all have English as their first language? (Kiedaisch and Dinitz 2007).

Some scholars have studied writing centers outside of the United States. As I mentioned earlier, the *Writing Center Journal* has an upcoming issue on the transatlantic writing center, and many other scholars have published research on writing centers outside the United States. However, a global perspective is still the exception rather than the norm.[1]

Harry C. Denny and colleagues (2018) worry that writing centers that continue to ignore their social obligations are failing to prepare the students who enter their writing centers to be global citizens. He and his coauthors continue this argument in *Out in the Center*, suggesting that writing centers must exist "as a social and cultural creation that extends into the world" (2018, 130). And this work is important because as Randall W. Monty (2016) explains, "In recognizing and naming what we are doing, we will also be able to identify the gaps and fissures of what we are not" (577). And we are not effectively looking outside North America or the Western tradition, or much beyond the nineteenth century when we tell the story of the writing center. I hope to establish that the tradition of the writing center is both ancient and global. As for which came first, the writing center or the writing classroom, I would say neither, as both are part of the same whole, the same circle, and a circle has neither beginning nor end. Consequently, our discipline has no reason to privilege either the writing center or the writing classroom.

What Was Taught at the *Edubba*?

Cuneiform was originally a logographic system, and well-suited for accounting purposes; however, as scribes needed new words to describe more abstract concepts such as names, or geographic locations, they combined existing signs to create syllables and used them as phonograms. This combining of a logographic system with a phonographic system "made it necessary for scribes to undergo a long difficult training" (Oppenheim 1964, 236).

Modern composition scholars use the term "translingualism" "as one possible entry point to contest the monolingualism that continues to dominate the teaching and study of college writing and reading in the United States and elsewhere" (Lu and Horner 2016, 207). And conversations about how to implement translingualism are occurring; however, a truly translingual approach has yet to be adopted by most writing programs, centers, and teachers.[2] We understand that today's writers must learn to write for an ever-expanding audience that includes people from across the globe who speak different languages and hold different beliefs. And we realize that it is the writing center's job to help them. Writing centers are "cultural and interdisciplinary contact zones" (Monty 2016, 420). But tutors in today's writing centers, regardless of where they are physically located, typically speak and write in one, maybe two languages. Many of the students who walk into our centers speak multiple languages, and many are not first speakers of the language spoken in that center. However, no writing center tutor would ever be told they were not eligible to work in a writing center until they learned to speak at least one other language fluently. They are only expected to be knowledgeable in the language privileged in that center as determined by the country in which that center is located. Many languages are spoken by the people of the United States; yet writing center tutors working in the United States are only expected to know English—and only standard English. Other countries privilege their national language in the same way. Some of this lack is caused by a dearth of research and pedagogical models (Canagarajah 2016), but the problem does exist. The opposite was true in the *edubba* where a translingual model was the norm. According to Oppenheim, the "bilingual literary tradition [of the *edubba* (Sumerian and Akkadian) [was] diffused throughout the entire Near East: from Elam and the Bahrain Islands (in the Persian Gulf) to the Hittite kingdom in central Asia Minor, to the west (between the Euphrates and the Mediterranean coast), and to Cyprus and Egypt" (13–14).

Despite its complexity, the educational process within the *edubba* was fairly prescriptive and progressed in phases. The type of tablet on which a student wrote, and the type of document they wrote on that tablet, provide key insights:

- Type 1. Large four-column tablets covered with long sections of text with no extra material. These tablets often contain a complete literary text (see figure 4.4 in table 4.3).

- Type 2. Smaller two-column tablets. The instructor's model fills the left column, and the student's copy of the model, the right. The text is normally ten to twenty lines long. The reverse side is covered in more excerpts in a left-to-right multicolumn format. Content includes lexical lists, syllabaries, and lists of personal names (see figure 4.3 in table 4.3).

- Type 3. Small one column tablets. These tablets often contain a complete excerpt from a larger literary text (see figure 4.2 in table 4.3).

- Type 4. Small lentil-shaped tablets with one column. The instructor would copy one or two lines, and below the student would copy them. The consistently poor quality on recovered lentil-shaped tablets suggests they were used for the most elementary exercises (see figure 4.1 in table 4.3). (Delnero 2010)

Several characteristics distinguish the work of the novice from that of the more advanced student and provide insight into the instructional process:

- Tablets produced by developmental students were the smallest in size and were made by flattening a fist-sized piece of clay.

The students' first work would be attempts to copy what the teacher had already written. I imagine that this exercise was very similar to the activity common in first-grade classrooms: copying the letters by following the letters printed on the top of the specially produced worksheet.

- While many type-two tablets have been recovered, on most tablets, the instructor's work is better preserved than the student's. These tablets show that a student's work was often scraped off so the student could try again. Multiple scrapings gradually reduced the thickness of that portion of the tablet, making it much more fragile.

- As students' skills developed, their work became more their own, and less or no text was written by the teacher either to be used as a model or to make corrections.

- Most of the existing literary texts that have been recovered survive because they were regular parts of the curriculum and were copied in abundance for decades, if not longer. Some examples include *Gilgamesh and Huwawa*, *The Song of the Hoe*, and *The Exultation of Inana*. (Delnero 2010, 59). This practice of selecting specific texts to use as models has been a common practice since this time and has obviously affected what literary works have survived.

Phase 1—Learn and Practice the Written Symbols of Cuneiform

The primary goal of phase one was to familiarize students with how to use a stylus to create the wedge shapes that they would combine to make characters. In addition to writing characters, the students would chant series of sounds (too-tah-tee) to learn the sounds associated with the symbols. During phase one, students produced type two and four tablets. Once they were proficient in these skills, they moved to phase two.

Phase 2—Citizen Training

In this phase, scribal students learned more than the craft of writing. They also received citizen training "through hymns, myths, and law codes" (Taylor and Finkel 2015, 27). During this phase, students produced type one and three tablets. Students began by writing people's names, including their own. As the proverb below, which was commonly copied by student scribes, makes clear, the inability to write your own name was a source of shame. "You are a scribe and you do not know your own name? Shame (?) on you!" (Black et al., n.d.). Students then advanced to writing and memorizing lexical lists, such as this one in table 4.1, which is a list of professions:

Students had to master specific technical terms associated with different professions (Pearce 1995): surveying, public administration, the law, commerce, the temple, medicine, astrology, etc. in order to learn the importance of audience: "The student was supposed to know the 'tongue' of several classes of priests, silversmiths, jewelers, shepherds, and master shippers" (Sjoberg 1975, 166).

This knowledge was essential to the scribes' final phase of education—the apprenticeship. Scribes were normally apprenticed to a family member to work in the family business or perhaps in an administrative position. To do this work effectively, the scribe needed to have memorized the Cuneiform symbols specific to that specialization.

Table 4.1. Lexical List of Professions—Used for Scribal Education

Text	Word	Translation
o ii 1' [. . .]	⌈šar-ra-nu⌉	[. . .] = kings
o ii 2' [. . .]	šar-ru a-šib pa-rak-⌈ki⌉	[. . .] = king who sits on a dais
o ii 3' [dumu] ⌈lugal⌉	DUMU šar-ru	son of a king = son of a king
o ii 4' [. . .]	mar-ti MIN<(šar-ru)>	[. . .] = daughter of a king
o ii 5' [. . .]	mar ru-be₂-e	[. . .] = son of a nobleman
o ii 6' [. . .]	mar kab-ti	[. . .] = son of a dignitary
o ii 7' [. . .]	be-lu-⌈tu₂!⌉	[. . .] = lordship

Source: Besnier, Marie-Françoise. 2014. The Geography of Knowledge Project. University of Pennsylvania. Accessed July 2021. http://oracc.museum.upenn.edu/cams/gkab/P338688.

For further practice, and to inculcate students into the scribal community, students copied proverbs such as these:

2.44 | You are an outstanding scribe; you are no lowly man.

2.47 | What kind of a scribe is a scribe who does not know Sumerian?

2.71 | Tell a lie and then tell the truth: it will be considered a lie.

(Black et al., n.d.)

It is not by chance that students spent a great deal of time copying proverbs, myths, hymns, and legends. Charpin believes scribal training was not just technical. "Fidelity to a certain notion of kingship has to be inculcated in scribes from the earliest years" to create esprit de corps among students who spend their "lives in the king's service" directly or indirectly (2010a, 45) as this excerpt from a hymn demonstrates: "The glory of your kingship will be in (everyone's) mouth in the edubba" (Sjoberg 1975, 171).

Today, writing centers continue to create esprit de corps among students to welcome them into the community of scholars and writers, and in the case of second language speakers, into their new culture. Elizabeth Boquet explains, "Community grows within these less structured, more social settings" (2002, 28–29), something the Sumerians discovered long ago.

Phase 3—The Study of Literature

At this point in their education, scribes began copying more complex works and sometimes began to write independently on style-one tablets. Students would have copied hymns in praise of the king, religious liturgies, poems, stories, and debates—a continuation of their citizen training. One of the reasons so many copies of *Gilgamesh* exist is its consistent use as an educational tool. Students also began to write about their scribal training, and it is these documents that give us the greatest insight into the educational practices. In the following document (see table 4.2), a supervisor offers advice to a younger student. Because of the document's length, I have used excerpts.

Table 4.2. Analysis

Excerpt	Analysis
9–15 "I just did whatever he outlined for me—everything was always in its place. Only a fool would have deviated from his instructions. He guided my hand on the clay and kept me on the right path. He made me eloquent with words and gave me advice. He focused my eyes on the rules which guide a man with a task: zeal is proper for a task, time-wasting is taboo; anyone who wastes time on his task is neglecting his task."	These lines clearly describe the writing instructor's practices. He • guides • advises • focuses These are the same practices used by modern writing center tutors. And anyone who has worked in a center can empathize with the teacher's difficulty in keeping his student from wasting their time.
16–20 "He did not vaunt his knowledge: his words were modest. If he had vaunted his knowledge, people would have frowned. Do not waste time, do not rest at night—get on with that work! Do not reject the pleasurable company of a mentor or his assistant: once you have come into contact with such great brains, you will make your own words more worthy."	Mesopotamian writing instructors understood the importance of not using their education to intimidate those they would teach. Modern studies have also shown that tutors who condescend or use exclusionary language, risk alienating their students.*
62–72 "Nisaba (goddess of writing) has placed in your hand the honour of being a teacher. {For her, the fate determined for you will be changed and so you will be generously blessed} {(1 ms. has instead:) You were created by Nisaba! May you …… upwards}. May she bless you with a joyous heart and free you from all despondency. …… at whatever is in the school, the place of learning. The majesty of Nisaba …… silence. For your sweet songs even the cowherds will strive gloriously. For your sweet songs I too shall strive and shall ……. They should recognise that you are a practitioner (?) of wisdom. The little fellows should enjoy like beer the sweetness of decorous words: experts bring light to dark places, they bring it to cul-de-sac and streets."	We all know that sometimes working with students can feel like a thankless job, yet this scribe's supervisor describes him as a "practitioner of wisdom," and believes students should enjoy their teachers' words as though they were beer! Praise teachers who bring light to dark places! Perhaps this attitude could be adopted by our modern-day supervisors and students.

Source: Created by the author.

*Raign 2013.

Like writing centers, an *edubba* could be located anywhere. The rooms identified by archaeologists as classrooms do "not necessarily contain distinctive furnishing or equipment of any kind beyond rudimentary bins for new and recycled clay" (George 2005, 7). According to Paul Delnero, early claims by archeologists to have found "classrooms" containing rows of benches have been widely disclaimed. Scholars now believe scribes were "trained in extremely small groups (1–3 students maximum) in the courtyards of private houses by a master (the *ummia*) who was often, but not always, the father or relative of the pupils, and not in an institutional setting like a school" (Delnero, email to author). A. R. George also believes that because students needed light to work by, instruction often took place outside rather than in the house itself. And Eleanor Robson, based on her excavation of an *edubba* in Nippur, describes the physical environment of the school as "much smaller and more domestic than the Sumerian school literature would have us believe" (2001, 62)—only about ten feet square. Clearly today's writing centers more closely resemble the *edubba* than the writing classroom. Rather than being a room full of rows of desks where a teacher lectures from a position of authority, both the *edubba* and the writing center favor looser configurations that allow students to write and collaborate more freely. In fact, Charpin believes that the term "school" is misleading because it "confers an institutional character on the training of young scribes that" didn't exist (2010a, 32). Students did not receive a commercial education in a large, structured classroom.

Who Taught? The Ancient Teaching Staff

> Regardless of their positions or backgrounds, our participants [12 writing center directors] all do everyday (day-to-day, administrative tasks), emotional (building and sustaining relationships), and disciplinary (engaging with/in the academic field) labor.
>
> —Caswell, Grutsch, and Jackson,
> *The Working Lives of New Writing Center Directors*

Although most *edubbas* taught only three to four students at one time, they normally had two instructors:

- The *ummia*, or expert
- The big brother

The *ummia* was also called the "school father," which suggests the nature of his relationship to his students.

The Ummia

While male teachers were the norm, women *ummia*, or "school mothers," did exist. The *ummia* received payment for teaching: sometimes they were paid directly in silver, sometimes paid in kind (clothing or food), and sometimes a student paid their way by working for the *ummia* after finishing their training (Cohen and Kedar 2011). Based on interviews with multiple writing center directors from disparate locations, Nicole Caswell, Jackie Grutsch, and Rebecca Jackson created a list of the writing center director's common duties:

- hiring and recruiting tutors;
- training tutors;
- mentoring team members;
- mediating conflicts, disciplining/dismissing team members;
- establishing or changing writing center policies;
- establishing new initiatives, changing or dropping existing initiatives;
- marketing the writing center;
- consulting with faculty, educating faculty about the writing center, working with faculty on writing assignments/assessments;
- attending university/school meetings and/or serving on committees;
- creating and executing assessment plans;
- attending and/or presenting at academic conferences, taking team members to conferences. (2016, 172–73)

Although they wrote this list in 2016, it is easy to imagine the *ummia* doing most of (if not all) these duties. Because they were paid, we must assume that *ummias* would have found it necessary to directly seek students—students provided their living. The average *edubba* only taught about three to four students, and often one or more of those students were the *ummia's* own children. However, to make a living, an *ummia* would have had to replace students who finished their education and build an initial pool of paying students—those who were not their children. Advertising for students could have taken several forms. It seems likely the *ummia* would have relied heavily on word of mouth, which means that they would have needed to do each of the duties listed above.

The *ummias* wrote about their experiences as teachers, which allowed them to provide each other with opportunities for faculty development and the exchange of teaching. Yoram Cohen and Sivan Kedar (2011) found documents written by teachers that they describe as "scholarly materials, relevant to their education in the scribal school" (223). This is lucky, because although thousands of student exercise tablets from the second half of the second millennium have been recovered, they tell us nothing about the teaching methods used in these schools. Several essays about school life like the one below also provide insight (Kramer 1981).

This passage from "School Days," a Sumerian document thought to have been written by an *ummia* around 2000 BCE, offers a funny satirical glimpse into life in the *edubba*. In fact, anyone who has ever taught likely has similar stories to share. In this piece, the *ummia* explains what happened after he chastised a student for multiple infractions—the student was caned several times. Clearly, the student had a different perception of his ability than his *ummia*:

> I spoke to me father of my hand copies, then
> read the tablet to him, (and) my father was pleased;
> truly I found favor with my father. (Kramer 1949, 205)

Feeling a bit too confident, the student takes some liberties the next day:

> In the tablet-house, the monitor said to me "Why are you late?
>   I was afraid, my heart beat fast.
> I entered before my teacher, took (my) place.
> My "school-father" read my tablet to me,
> (said) "The . . . is cut off," caned me. (205)

The student's day continues along these lines, and he goes home and tells his father of his problem. The father immediately invites the teacher to dinner and treats him as an honored guest and praises his teaching skills:

> The teacher was brought from school;
> having entered the house, he was seated in the seat of honor.
> The schoolboy took the . . . , sat down before him;
> whatever he had learned of the scribal art,
> he unfolded to his father.
> His father, with joyful heart
> says joyfully to his "school-father":
> You "open the hand" of my young one, you make of him an expert,
> show him all the fine points of the scribal art. (206)

After praising the teacher, the father gives him wine, oil, a new garment, and a band for his hand. The teacher responds with praise for his student:

> Young man, because you did not *neglect* my word, did not forsake it,
> May you reach the pinnacle of the scribal art, achieve it completely.
> Because you gave me that which you were by no means obliged (to give),
> you presented me with a gift over and above my earnings, have shown me great honor,
> may Nidaba, the queen of the guardian deities, be your guardian deity,
> may she show favor to your *fashioned* reed,
> may she take all evil *from* your *hand copies*.
> Of your brothers, may you be their leader,
> of your companions, may you be their chief,
> may you rank the highest of (all) the schoolboys,
> . . . . who *come* from the royal house.
> Young man, you "know" a father, I am second to him,
> I will give speech to you, will decree (your) fate:
> Verily your father and [*mother*] will support you *in this matter*,
> as [*that*] which is Nidaba's, as that which is thy god's, *they* will present offerings and prayers to her;

> the teacher, as that which is your father's verily will pay homage to you;
> in the . . . of the teacher, in the . . . of the big brother,
> your . . . . whom you have established, your manly [kinfolk] verily will show you favor.
> You *have carried out well the school duties*, have become a man of *learning*.
> Nidaba, the queen of the place of *learning*, you have exalted.
> O Nidaba, praise! (205–6)

Perhaps it was common for *ummia* to accept gifts (Alster 1975); we have no way of knowing. Perhaps the writer of this document, which was copied by innumerable schoolboys, was sending a message to both his students and their parents. Perhaps he was simply offering a humorous glimpse into his life as a teacher. And while we do not cane students, or accept gifts from students or their parents, it is funny (and reassuring) to recognize and appreciate another teacher's frustration and his willingness to reverse his opinion of a student who is willing to learn. This story also demonstrates that even the first writing center was "neither a haven for togetherness nor an island of intimacy insulated from political and social relations" (Geller, et al. 2007, 7).

Ummias did not teach alone. They were assisted by the big brother/sister—an older more experienced student. The *ummia*, like his modern counterpart, would have trained and supervised the big brother/sister. Again, though few, big sisters were also a presence in the *edubba*.

The Big Brother/Sister

The big brother/sister prepared "discipline-specific tutorials" (Brauer 2002, 71) for students, reviewed their work with them, and listened as they recited their lessons (Kramer 1981). Because the tablets from which students would have studied were largely lists of words and terms, Samuel Kramer suggests that the big brother/sister would have supplemented the *ummia*'s lectures with oral instructions. Sadly, because they were oral, we have no record of these conversations. We do, however, have brief texts such as this one, which illustrate that relations between tutors and students have changed very little. A scribal student makes the following claim to the big brother:

> I really know my scribal knowledge,
> I don't get stuck at anything!
> My teacher shows me a certain sign,
> I add one or two more from memory!
> Now I've been here for the stipulated time
> I can cope with Sumerian, scribal work, archiving, accounting, calculation!
> I can even hold a conversation in Sumerian! (George 2005, 2)

After listening to the student boast of his skills, the big brother responds, "If that is so, Sumerian must be keeping its secrets from you" (George 2005, 2). How often have we listened as a student tried to boast of their expertise to convince a tutor that they needed no help? Apparently even ancient tutors could tell from the quality of a student's work when that student was boasting.

In the following passage a big brother vents his frustration on a student who failed to do his work, and then brags about how his own hard work makes him superior to his student.

> "Well, fellow student, what shall we write today on the back of our tablet?"
>
> "Today we will not even write a single word from our lesson!"
>
> "But then surely the teacher will know and be angry with us because of you; what will we say to him?"
>
> (The senior student then brags of his superiority)
>
> ". . . I have become excellent in the scribal art; I have fulfilled the function of 'big brother' to perfection! You are slow of understanding and hard of hearing; you are but a novice in the school! . . . My father speaks Sumerian; I am the son of a scribe; but you are the son of a vile one, a barbarian; you cannot shape a tablet, nor knead an exercise tablet. You cannot even write your own name . . . .'
>
> [After a missing section, it appears that the teacher arrives and intervenes.]
>
> Teacher: 'Why do you act like this? Why do you push one another and hurl insults at one another? You raise a clamour in the school! . . . . (Vantisphout 1997, 589–90)

Unfortunately, most of us have had experiences with tutors such as this young man who blamed a student's failure on the student rather than considering that his tutoring might be to blame. Human nature is reassuringly constant. As I stated previously, though evidence of women as both instructors and students is less common, they did exist.

WOMEN TEACHERS

While textual evidence of both male teachers and students is much more prevalent, we do have proof that women both taught and attended school. According to Samuel Meier, males dominated in the communication network, but women were not locked out. One type of textual evidence for women scribes can be found in the period's literature. The first patron of scribes, who oversaw their craft, was the goddess Nidaba, also called the queen of scribes:

> On the day, the shepherd-man went out to the plain.
> Where, oh Dumuzi (are you going)?—I will go to the stall.
> His sister, Queen of scribes,
> Was standing (?) there (?) in the open air (?) (Alster 1975, 23–25)

However, despite this clear evidence of women as scribes in the mythology of the culture, evidence of actual woman scribes has been hard to locate. According to Laurie Pearce (1995), the first verified woman scribe was Enheduanna, daughter of Sargon of Akkad, thought to be the author of a Sumerian poem *In Praise of Inanna* (Pearce 1995, 2266). Another woman, Nishatapada identified herself as a trained scribe in her letter to the king of Larsa (Tetlow 2004) and her elegant prose "became the object of study in subsequent generations of apprentice scribes" (Pearce 1995, 2266).

In 1962, Rivkah Harris identified evidence of eight naditu women[3] (1962, 1) who worked as scribes during the reign of Hammurabi, including Inanna-amamu, who was a naditu *and* a scribe. Her father Abba-tabum was also a scribe, which suggests that he would have trained her in his *edubba* (8). Meier argues that the number of women scribes proved to have existed suggests that the "educational investment in women was not on a small scale" (1991, 542). It is also likely that naditu scribes taught other naditu women in their homes within the naditu community following the tradition of the *edubba* (George 2005). One final piece of textual evidence points to naditu scribes training other scribes. A study of contracts from Sippar found a recurring phrase "At the opening of the lattice" (Harris 1964, 130). Some scholars took this as evidence that the naditus of Sippar lived a cloistered life. However, later studies suggest "its use was probably only a scribal fad without any real significance" (130). For a scholarly vogue to have developed among a specific group of naditu scribes, it had to have been an off-shoot of their scribal training. Clearly women, though they had less access to education than men, did learn to write and work as both *ummias* and scribes.

What Methods Did They Use?

Scribal education was lengthy, and students would have worked with both the school father/mother and the big brother/sister as they progressed through the stages of education—all at the same *edubba*. In effect, the *edubba* was the source of primary, secondary, and higher education, and as students gained knowledge their role in the school would have shifted. It seems likely that most students would begin as simple students, progress to the role of big brother/sister, and eventually become either a professional scribe or an *ummia*. Students began by learning to make simple marks on clay, then moved to copying and memorizing lists of syllables, then words. Once students had learned an adequate number of words, they began learning to combine them into sentences. During the final stages of their education, students copied the great works of literature, and began learning specialized vocabularies and forms of writing. They also copied letters from a corpus of about thirty exemplary letters collected by schoolmasters between 2001 and 1000 BCE. This activity, more than being encouraged as a form of imitation, was meant to provide scribes with practice in correctly writing Cuneiform and as a form of cultural indoctrination: because many of the letters were written by scribes serving the king, apprentice scribes became familiar with the ruling powers to whom they owed allegiance. What is most interesting is the clear evidence that students collaborated as they copied their assigned texts.

An Emphasis on Collaboration

To learn how to write Cuneiform, students spent hours copying texts written by either the *ummia* or the big brother/sister. As the texts they copied became more complex, students began to make intentional changes to the original documents, and because they collaborated, they influenced each other's work. C. Jay Cristomos bases this judgment on the reoccurrence of unique words used in texts found at the same site. He identified three different texts used in scribal education in the Old Babylonian period that include lemmata (especially rare words):

1. the word list of Izi and the so-called Enheduana corpus
2. the sign list Ea and Sumerian Proverb collections
3. the world list Lu-azlag and the Eduba text and dialogues. (2015, 133)

One example Cristomos cites from the Sumerian Proverbs collections states: "A scribe without a hand, a singer without a throat" (135). Cristomos believes that the sign used for "throat" in this quote does not appear anywhere else in Sumerian literature. Further evidence, he argues, that the collection of texts containing these lemmata were written by scribal students from the same *edubba* who were influencing each other's word choices. This sort of collaboration is still encouraged today, and most writing center staff would argue that they can easily tell which students have worked with which tutor because those students tend to repeat words and word combinations privileged by that person.

Delnero also argues that teaching in the *edubba* was highly collaborative. Based on his identification of repeated errors and anomalies in two Old Babylonian literary tablets, he contends that students imitated and were influenced by their fellow students: "While it is not impossible that [two] scribes made the same memory error . . . the error . . . may have been present in the model text or occurred at an earlier stage when the two scribes were rehearsing the text they were learning together" (2010, 66). As I mentioned earlier, another occurrence of such a trend is found in texts recovered from a naditu enclosure. A study of contracts from Sippar, found a reoccurrence of the phrase "At the opening of the lattice" (Harris 1964, 130). The *edubba* also focused on teaching specific genres and formulas for writing within those genres.

An Emphasis on Genre and Form

Because they were being trained to work as professional scribes, students of the *edubba* were taught both professional terminology and professional genres. For instance, students learned how to write letters, legal documents, recipes, and rituals. When writing manuals of recipes and ritual incantations, students were taught to use conditional phrases, second person, and step-by-step instructions (Halton and Svard 2018). Because letters were the primary form of communication, they made up the most important genre (Charpin 2010a). Students were taught to write two specific types of letters.

Letters written in the Sumerian to Babylonian Period 1600–600 BCE. This form of letter was used for giving administrative instructions. Students were taught to use a standard greeting, and the instructions were always given in imperative voice (Oppenheim 1964, 277).

The Old Babylonian Period On. Letters from this period used a highly elaborate greeting that included numerous blessings based on the relationship between the sender and the receiver—further indication of the emphasis given to the rhetorical situation within the edubba (Oppenheim 1964, 277). However, "the discipline they impose on the wording is normally in evidence at the beginning of the text (introductory and salutation formulas) and affects the body or the end of the letter much more rarely" (Oppenheim 1967, 64):

> Tell the Lady Zinû: Iddin-Sin sends the following message:
>   May the gods Šamaš, Marduk, and Ilabrat keep you forever in good health for my sake. (84)
>
> Tell my father, whom the gods, Gula, [. . .]. Damu, and Urma-šum keep in good health: Warad-Gula sends the following message (85)

As the two greetings demonstrate, a well-schooled scribe knew to include both a blessing from a god or gods and to clearly state the name of the person sending the message. Though the order of this information could vary, both elements were required. Even women slaves were known to send letters, though they would have likely hired a scribe to write the message. As this greeting suggests, the scribe hired by this slave was not as qualified as others and forgot to include the blessing—an important oversight given that the slave is making a request of her master: "Tell my master: Your slave girl Dabītum sends the following message" (85). And though the structure of a letter might have been less rigid than that of a legal document, a scribe was still expected to demonstrate a knowledge of the proper format: "What is this behavior? Even when I write by the rules, you don't send me any reply to my letter!" (Charpin 2010a, 96). How often have you heard modern students make a similar complaint? "I followed the rules and did exactly what my instructor said, but I still didn't get an A!" Perhaps this student should have paid attention when they were told to revise, a practice heavily emphasized in the *edubba*.

AN EMPHASIS ON REVISION

Many texts, written between 3000 and 1600 BCE have been recovered from across the ancient Near East (Western Asia, Turkey, and Egypt) (Charpin 2010a). The oldest texts were recovered from Uruk, Larsa, Tell Uqair, and Jemdet Nasr. The largest body of educational texts comes from the Old Babylonian period (2000–1600). Table 4.3 on the next page shows examples of the types of student tablets:

Which is More Like the Edubba? A Composition Course or a Modern Writing Center?

The composition classroom is a place of organized transmittal of information from expert to novice in a controlled setting that allows for limited interaction between student and teacher because the teacher is drastically outnumbered by the volume of students. A writing center, by contrast, is a place "where small groups of people often work together quite closely for several years" (Boquet 2002, 28–29)—a description that aptly suits the *edubba* where more than three to four students were rarely taught at one time. Furthermore, these students often studied together for years (Delnero 2019) and were tutored individually: and one truth that all writing center directors accept is that "a writing center is not a writing center without one-to one tutoring" (McKinney 2013, 58). In the e*dubba*, the *ummia* and the big brother/sister did not lecture groups of students. Instead, they worked with them individually.

Textual evidence suggests that the teaching methods used today in our writing centers were influenced by the practices developed in the *edubba*. Carino describes writing centers as . . . "places where writing [is] more likely to be viewed as a process, where staff would be reconceiving notions of pedagogy according to this new paradigm of composition stud-

Table 4.3. Types of School Tablets

Type of Tablet	Purpose	Illustration
Small Lentil-Shaped Tablet	Normally included one or two lines of a particular text in well-crafted script (teacher), followed by a copy of the lines in poor to excellent quality (student). The normally poor quality indicates that lentils were used early in a scribe's training.	Figure 4.1. Lentil-shaped tablet. Source: Metropolitan Museum of Art, public domain.
Small One-Column Tablets	Usually one extract from a longer literary work. Inscribed on obverse and reverse. Usually 20–50 lines.	Figure 4.2. Small one-column Cuneiform tablet: record of rations of beer, bread, oil, and onions for messenger. Source: Metropolitan Museum of Art, Public domain.

Type of Tablet	Purpose	Illustration
Two-Column Tablets	Left is the instructor's model; right is the student's work. Used during the elementary phase of scribal education.	Figure 4.3. Cuneiform tablet: Late Babylonian grammatical text. Source: Metropolitan Museum of Art, Public domain.
Large Tablets, Cylinders, or Prisms	Divided into four columns. Used to produce an entire literary work. Used in the advanced stage of training. (Delnero 2010)	Figure 4.4. Cuneiform cylinder with inscription of Nebuchadnezzar II. Source: Metropolitan Museum of Art, Public domain.

Source: Created by the author.

ies, where people found a place to experiment, to pose questions, and to seek solutions'" (1996, 35). Gerd Brauer describes the essence of such a place as writing-centric, and offers the Freiberg model, which developed in Germany, as a method of involving both writing center tutors and faculty in the creation of writing-centric curriculums that cross the bounds of the classroom—another reason to question why we artificially separate the writing center from the classroom, giving the classroom precedence.

Conclusion

If we have the imagination to envision an interdisciplinary history for our discipline, a history not tied to the Western tradition or the English language, we can imagine students and mentors working together, collaborating, talking, and laughing as they learned to write on clay tablets. It is easy to use the past as a justification for congratulating ourselves in the present: as in, *look how much better we've made things*. But perhaps ancient writing instructors have something to teach us. And rather than looking at writing centers and writing classrooms as competing developments on a linear timeline, perhaps we should remember they are parts of one whole, equally important to the students they serve. And perhaps those of us who study writing centers should see our scholarship differently as well. Perhaps our scholarship should not be described as a horizontal timeline but rather as a circle with no beginning or end in which knowledge flows equally in both directions. Within this loop of information, there is no room for top, bottom, first, or last.

In service of this goal, I suggest a new travel itinerary. Let's throw out the map with its boundaries and explore every aspect of our rich history openly and equally, and without regard for power structures, like that annoying voice on our GPS that tells us to turn right when we really want to go left.

Chapter 5

Mythos, Nomos, Logos

Evidence of Sophistic Reasoning before the Sophists

Hey! Hoe, Hoe, Hoe, tied up with string;
Hoe, made from poplar, with tooth of ash;
Hoe, made from tamarisk, with a tooth of Sea thorn;
Hoe, double-toothed, four-toothed;
(5) Hoe, child of the poor, *bereft even of a loin-cloth*;

Hoe picked a quarrel with Plow.
Hoe and Plow—this is their dispute.

—Vanstiphout, *Dispute Poems and Dialogues
in the Ancient and Mediaeval Near East*

You are standing in the courtyard of your school. The sun beats down on your head, and sweat drips into your eyes. Opposite you, one of your classmates stands—legs braced, arms crossed—ready for battle. Your ummia, school father, tells you this. You are night. You are superior today. He turns to your opponent. You are day. What is night to you? Next, he speaks these words: "Night, start a wrangle." Memories of the disputations you have studied flood your brain as you seek for your opening words: Day, what are you to me? I provide relief from your heat, so that all might rest peacefully. You cook the grains of wheat in the shaft. Day, what are you to me?

Intentional acts of persuasion developed in tandem with humans' need to communicate. At the most basic level, even a smile is a form of persuasion (approach me), as is a snarl (stay away). However, intentional acts of persuasion, as well as the teaching of those acts, is almost universally attributed to the Greeks. First the sophists are recognized for commercializing the teaching of persuasive speech. Then Aristotle, a systematic rhetorician, divided the speech into four parts: the introduction, statement of the case, proof of the case, and the conclusion. Next, the author of the *Ad Herennium* added two more parts: outline of the points or steps in the argument, and the refutation of opposing arguments, while Plato turned the act of persuasion into a philosophical search for truth that relied on the use of the dialogue. But in fact, the ancient Mesopotamians had already done all of this in their disputations.

The disputations are a fascinating mix of what would later become clashing ideologies: they use the form of a dialogue to explore a question, but the truth that is always revealed at the end is based on a string of probabilities, and one truth itself is often rejected in favor of another. And though the disputations are written as dialogues, they follow a formulaic structure of introduction, argument, disputation, and conclusion. But what is most significant is the disputations' use of both rhetorical and sophistic principles. While the disputations make heavy use of systematic rhetoric's principles of logos, pathos, and ethos as well as the traditional structure of the persuasive speech, they make equal use of the sophistic concepts of mythos, nomos, and logos. In a sense, it's as though the Mesopotamians experimented with and combined what would later become distinct rhetorical schools of thought.

The Structure of the Disputation

The disputations are structured as a dialogue between two speakers, making them very similar to the Platonic dialogues, though the Platonic dialogues do not use a formulaic structure. The disputations, on the other hand, begin with a prologue that states the beginning of the argument, include a formal debate in dialogue form between two participants, and end with the adjudication—naming of a winner. In effect, the disputations' structure combines the dialogue format of Plato, with the systematic structure of the Greeks.

PROLOGUE/INTRODUCTION

The Mesopotamian "Prologue" and the Greek "Introduction" have two primary purposes:

- to state the nature of the argument
- to render the audience receptive to the argument(s)

Both the Mesopotamians and the Greeks relied on the use of ethos to establish the credibility of the speaker as a means of appealing to the audience. The Mesopotamians developed ethos by using lengthy descriptions of the cosmological origins of the two speakers to demonstrate their qualifications, while the Greeks touted their personal intellectual and moral qualifications to win the audience's trust. In this prologue, Bird and Fish build their ethos from the fact that Nudimmud, a god, created them, let them multiply, and taught them their place in the world:

> Nudimmud, the noble prince, the lord of broad insight,
> When he had fashioned Bird and Fish,
> He filled canebrake and marsh with Fish and Bird,
> (20) Selected their stations,
> And made them acquainted with their rules. (Vanstiphout 2003b, 581)

In this excerpt of the introduction to "The Two Cultures," writer C. P. Snow describes his personal background to establish how he is qualified to compare the sciences to the humanities: "By training I was a scientist; by vocation I was a writer. That was all. It was a piece of luck, if you like, that arose through coming from a poor home" (Snow 1961, 1). Although the style of the two examples is obviously different, both rely on the same rhetorical tool, and both have the same purpose. I would argue that this suggests that what we still recognize today as the parts of a persuasive argument are based on their effectiveness rather than some arbitrary decision by the writers who first used them. Function proceeded form.

Formal Debate/Statement of Fact and Confirmation

The purpose of the Mesopotamian formal debate is the same as that of the statement of fact and the confirmation: to make and refute arguments. Again, because the disputations are written as a dialogue between two speakers, the making and refuting of arguments is directed at one's opponent in a very literal manner. However, within persuasive speeches such as *Encomium to Helen*, Gorgias made his own argument then disputed what he believed were the likely arguments to be raised by his audience, which achieves the same goal. In both instances, the same writer wrote both the argument and counterarguments because our audience is always a fiction to some degree (Ong 1975). Sound familiar, Plato?

ADJUDICATION/CONCLUSION

An effective persuasive speech doesn't just end. Instead, the writer uses the ending to reinforce their argument and reiterate the point they wish the audience to remember: "Like a maid-servant always ready, you will fulfill your task! Because the hoe was greater than the plow, praise be to Nisaba" (Vanstiphout, 2003c, 580–81). In this example, the adjudication serves to tell the reader that the hoe was the winner of the argument because of its humble and useful nature. The plow is equally useful, but only after hoe has done the initial work. In other words, the mighty depend upon the humble. In the conclusion of *Encomium to Helen*, Gorgias reiterates his primary argument: "I have by means of a speech removed disgrace from a woman. I have observed the procedure which I set up at the beginning of the speech; I have tried to end the injustice of blame and the ignorance of opinion" (Gorgias 1990, 42). In both cases, the truth discovered at the speech's end is the same truth predetermined by the writer. However, while the characters in Plato's dialogues masquerade as real people engaging in the actual process of debate, the archetypes that speak in the disputations do not pretend to represent real life nor "real" truth.

Beyond similarities in structure, the purpose of the disputations in Mesopotamian society demonstrates both their rhetorical function and their use of sophistic reasoning.

The Disputations' Purpose

Although they are often categorized as literature, the primary purpose of the disputations was to educate. Like the Homeric epics of the fifth century, the disputations shared "cultural instructions—nomoi and ethe" (Jarrat 1991, 33). According to Eric Havelock, both the structure and performance of the epics encouraged the audience to completely accept the cultural code being shared without question (1963). And within the disputations, the adjudication, a required element, is given by either a god, goddess, or king, ensuring the audience's acceptance of the decision made—both ethical appeals. For example, in *Bird and Fish*, Fish destroys Bird's nest because Fish is losing the argument. However, despite Fish's superior usefulness, Bird is declared the winner because violence can never be justified, and that message had to be taught.

Assyriologist H. L. J. Vanstiphout describes the disputations the following way (I have put significant words in italics):

> The genre may be said to consist of a frivolous, or at least *playful*, evasive exercise of matching wit against wit (or more precisely pretending to do so), and so reaching somewhat unexpected results, by *opposing almost anything*, by drawing the evidence from *common knowledge* (meaning: What everybody can see everyday must be true!"), by throwing in learning and erudition (meaning: Deeper, more serious knowledge, of course based upon scribal erudition, may change your naive beliefs!"), and basically by observing that *nothing is exactly what it seems*, and certainly that *all coins have two sides*. (1991b, 45)

While Vanstiphout acknowledges that the disputations are rhetorical, he certainly never suggests that the Sumerian disputations should be considered as precursors of sophistic rhetoric. But many of his terms echo those used by both the sophists and modern scholars of the sophists.

Protagoras wrote, "Of all things the measure is man" (1972, 18), echoing Vanstiphout's belief that within the disputations what is seen is what is true. Jasper Neel (1988) describes sophistic rhetoric as playful. Finally, Vanstiphout alludes to what many consider the sophist's creed with his reference to nothing being what it seems and all coins having two sides: "The assignment of a particular value depends on social and historical circumstances" (Bizzell and Herzberg 1990, 23).

In this chapter, I seek to demonstrate that while other scholars accurately use the generic term "rhetorical" when describing the disputations, they can more accurately be described as precursors of sophistic rhetoric. I will analyze one of the disputations, *Hoe and Plow*, to show that it relies heavily on the use of pathos and ethos, while also using both mythos, logos, and nomos to teach the rules of social order, social cohesion, social functioning (Ponchia 2007). However, we must start by understanding the function of the disputation.

What Did the Disputations Teach?

The disputations were intended to teach, but what they taught was twofold. The disputations were first a tool for scribal education, and next a tool for societal education: mythos to explain origins and technological discoveries, logos to teach structure, and nomos to teach behavior. Scholars agree that copying the disputations would have been part of the education of scribes as they completed their schooling. *Hoe and Plow* and *Ewe and Grain* were the most prolifically copied disputations. There are more than sixty manuscripts of *Hoe and Plow*, and approximately seventy of *Ewe and Grain*. Most manuscripts were from Nippur (Jimenez 2020, 15–16).

Certainly, the practice of copying the disputations would have helped scribal students to master the Cuneiform writing system. In fact, Vanstiphout (1991a) argues that parts of *Tree and Reed* are based rather simplistically on lexical lists, which scribes used to master the Cuneiform system:

> The Nurmagal tree, the Apple tree, the Vine . . . .
> The Lam tree, the Alanum tree the Popular,
> The Urzinum tree . . . (32)

This listing of types is sometimes replaced by a more complex listing of lexically related terms:

> You cannot dam up water when it escapes;
> You cannot heap up the earth in the basket;
> You cannot press clay to make bricks;
> You cannot lay foundations or build a house;
> You cannot strengthen an old wall's base; (32)

Rather than simply listing words without logically connecting them, the writer shows a logical progression: water, earth, brick, house. The writer creates a list with a parallel structure that links each deficiency of the Plow with the others. The repetition of the words "You cannot" emphasizes what the Plow cannot do, thus emphasizing what the Hoe can.

However, as Ponchia explains, the scribes were doing more than robotically copying manuscripts. She argues that the rhetorical skills the scribes were learning "provided [them with] the technical tools to cope with the most critical issues of society, fostering, at least, the process of interpretation of traditionally accepted mythical, social and political principles" (Sumer 2007, 78)—the same skills taught by the sophists.

Are the Disputations Rhetorical?

Throughout his series of three articles on the disputations, Vanstiphout refers to the use of rhetorical connections between arguments, the use of opposing arguments, and "the ubiquitous use of point-to-point reversal of arguments adduced by the opponent"(Vanstiphout 1992, 343). Vanstiphout is not the only scholar to discuss the rhetorical attributes of the disputations. William J. Hallo (2004) believes that Sumerian disputations have a strong claim to being "true exercises in rhetoric" (30). Ponchia argues that the personifications within the disputations make use of rhetorical topoi such as utility as a rhetorical strategy. And I argue that it is the undeniable similarities to the practices of the sophists that is key to understanding the rhetorical significance of the Sumerian disputations.

What Makes the Disputations a Precursor to Sophistic Rhetoric?

Rich Enos was the first to push back against the notion "that not only did the art or techne of rhetoric emerge in the fifth-century BCE Greece, but any examination prior to this period is 'irrelevant to the proper history of rhetoric'" (Enos 1993, 2). He argued that the Homeric epics contain examples of heuristic, eristic, and protreptic discourse suggesting they have "much to reveal about the epistemic development of rhetoric" (Enos 1993, 2).

Plato described eristic discourse as "disputatious, controversial, pugnacious, combative" wrangles (90–91), which are forms of sophistry and encouraged the practice of protreptic discourse, a form of rational inquiry "because it provides direction for thought leading to knowledge." Plato argued that the sophists practiced eristic discourse, which meant ignoring truth to win the argument. The philosophers, however, practiced protreptic discourse to uphold truth. The Sumerian disputations include examples of both eristic and protreptic discourse but not for the purposes that Plato supposes.

The disputations are eristic in the sense that they are wrangles. The word "wrangle" is used in many of the disputations to signal the beginning of the debate:

"Fish to Bird cried out;
[. . .] started a wrangle." (Vanstiphout 2003b, 581)

But eristic discourse was not used to allow one participant to win at any cost; instead, when combined with protreptic discourse, the dispute was intended to reveal a truth to its audience because the ultimate purpose of the disputations was to teach community-specific customs and laws. This reliance on nomos is another link to sophistic rhetoric.

Susan Jarratt describes the sophists as the first teachers of systematic instruction in "the art of speaking and writing" (Jarrat 1991, xv). For a price, the sophists would teach the young men of the emerging Greek democracy to participate effectively in political life, and their success was predicated upon their ability as persuasive public speakers. What separated the sophists from those who followed—Plato, Aristotle, Cicero, Quintilian—was their belief that "truth" was a fluid construct that "had to be adjusted to fit the ways of a particular audience in a certain time and with a certain set of beliefs and laws" (xv). Like the sophists, the *Ummia*, or scribal teacher, charged a fee to teach. The structure of the disputations also suggests that *Ummia* taught students that truth was flexible and based on the needs of the society. However, no other scholars have specifically identified the rhetorical principles found in the Sumerian disputations as sophistic, recognizing that what occurred in the fifth century was *evolutionary* rather than *revolutionary* (Swearingen 1986).

To address Jarratt's first claim, the Mesopotamians did systematically teach scribal students how to write persuasively. As discussed in previous chapters, they received this training as they learned to write transactional letters and instructions, and they expanded their knowledge of rhetorical practices as they copied the disputations and, perhaps, performed them. Finally, scribal students also worked with the *Sumerian Rhetoric Collections*, "collections of very diverse types of mostly fictive, non-casual discourse which were used to instruct non-native speakers of Sumerian in the rhetoric of Sumerian appropriate to educated and literate Old Babylonian Mesopotamians, primarily in the schools themselves" (Falkowitz 1980, 4).

Secondly, the disputations demonstrate the culture's belief that knowledge was fluid. For example, in several of the disputations, the more effective speaker is not declared the winner. Instead, the speaker who best serves society is the winner, and an emphasis is placed on the interdependence of the speakers, making them equally important in reality, if not within the disputation—an important lesson for a society heavily reliant on agriculture. As Ponchia argues, the disputations provide a "process of interpretation of traditionally accepted mythical, social and political principles" (2007, 78), which are naturally fluid because they are bound by societal context.

Vanstiphout argues that the disputations each contain a moral lesson, the teaching of the lesson being the primary function of the disputation:

- Hoe and Plow—The superiority of the commoner over the high and mighty (Vanstiphout 1992, 347)

- Bird and Fish—High moral virtue is no excuse for intolerance and violence (Vanstiphout 1992, 348).

Clearly, what Vanstiphout described as moral lessons can accurately be described as examples of appropriate social behavior, or nomos. And interpreting mythical, social, and political principles using rhetorical tools is precisely what the sophists did. However, the Sumerians might have practiced a nascent form of sophistic rhetoric. The fact that the Sumerians didn't codify the rhetorical strategies they employed doesn't mean they didn't consciously teach and use them. Specifically, an analysis of the disputations demonstrates that the writers used both ethos and pathos to make their argument, the primary function of which was to intermingle mythos, logos, and nomos to teach lessons in correct societal behavior.

Mythos, Logos, and Nomos

The emphasis on societal questions of behavior and values in the disputations foreshadows the practices of the sophists.

MYTHOS

A mythic era is defined by a culture's use of myths to explain "natural phenomena, detailed codes of everyday behavior, and even geographical and technological information" (Jarrat 1991, 32). In ancient Mesopotamia, the invention of writing is explained in the myth *Enmerkar and the Lord of Aratta*:

His speech was substantial, and its contents extensive. The messenger, whose mouth was heavy, was not able to repeat it. Because the messenger, whose mouth was tired, was not able to repeat it, the lord of Kulaba patted some clay and wrote the message as if on a tablet. Formerly, the writing of messages on clay was not established. Now, under that sun and on that day, it was indeed so. The lord of Kulaba inscribed the message like a tablet. (Vanstiphout 2003a, 85)

Within the disputations, all but one, the *Hoe and the Plow*, has a cosmological prologue based on recognized creation myths (Jimenez 2020). The myths contained in these prologues address both natural occurrences and technological developments:

Ewe and Wheat—before the gods made Ewe and Wheat, humans roamed naked and ate grass like sheep. Seeing this, Enki and Enlil sent humans Ewe and Wheat, and they brought prosperity and created opportunity for technological developments such as the making of thread, the process of weaving, the making of clothing, and the baking of bread.

Bird and Fish—Enki created a place for all—pens and stalls, cities and villages, and a king who "rose as daylight over the countries" (Vanstiphout 2003b, 581). He also created places for both Fish and Bird. To thrive, all must stay in their appointed place—fish don't belong on land, and birds don't belong in water. However, the breaking of these rules does not justify violence.

Summer and Winter—the gods created summer and winter, and both benefit humankind. Without the snow of winter, there would be no water in summer, just as without the farmer who feeds the people, there would be no king. (Vanstiphout 1991b)

Logos

Traditionally, scholars believed that preliterate societies were incapable of using logos because of the limitations placed on their ability to think logically within an oral framework, and without logos we have no argument.[1] The Homeric epics also contain rudimentary examples of logical argument. In fact, each of the Sumerian disputations has a logical structure, and while each disputation included the same parts, interesting modifications were made within the individual disputations to strengthen the argument based on the rhetorical context.

According to Enrique Jimenez, to be a disputation, the poem must meet these criteria:

1. Disputations are poetic and written in verse form

2. They include an introduction (or prologue), disputation, and adjudication

3. The disputation is mostly dialogue with only small occurrences of narrative

4. The two participants are articles or animals (i.e., Hoe and Plough/Fish and Bird)

5. The purpose of the dialogue is to establish the superiority of one participant over another (2020, 11–12)

Each of the six disputations that meet his criteria seek to establish which protagonist best serves humanity. Over the course of the disputation each participant speaks multiple times; sometimes they speak an equal number of lines, and sometimes they do not. But the disparity seems intentional. For example, in *Hoe and Plow*, Plow speaks only one-third as many lines as Hoe. However, this is an intentional rhetorical move on the part of the writer: as Hoe argues, it works for twelve months out of the year, while Plow works only four. Clearly, Plow is allowed to speak an amount equivalent to his work (Vanstiphout 1990), an example of social justice.

Nomos

The ultimate purpose of the disputations was to teach community-specific customs and laws. Like the sophists, the Mesopotamians used *dissoi logoi*[2] to make their arguments, but unlike the speeches of the sophists, the disputations always ended in a conclusion—one speaker was declared the winner. However, who won was not what mattered. What mattered was the importance of recognizing how their interdependence benefits society. What is the Hoe without the Plow? What is the Summer without the Winter? The Ewe without the Wheat, the Fish without the Bird? This example is from *Ewe and Wheat*:

> Thereupon Enki spoke to Enlil:
> (180) "Father Enlil, Ewe and Wheat, both of them,
> Should walk together!
> Of their combined metal [the alloy] should never
> Cease;
> Yet of these two Wheat would be greater! (Vanstiphout 1991b,
> 577)

As Ponchia explains,

> The ultimate aim of the confrontation is to specify, select and order values, confirming in this way those very assumptions that give fuel to the arguments, e.g.: prestige wins over factuality, usefulness in religious matters is nobler than usefulness in common ones, but the creation of abundance for the land and the protection of the humblest and poorest categories are the most appreciated virtues. Society is, like the world of the debates, made up of complementary elements, all of them useful, all of them productive in an ordered cosmos, all of them worthy of consideration and protection. (2007, 70)

In *Fish and Bird*, the moral lesson is simple: stepping outside your designated place in society is wrong but should never be met with violence. However, this simple truth reveals a deeper meaning. The gods gave each thing in the universe an appointed place. Some positions are humbler than others, and those in humble positions should remain in those positions and willingly serve those with higher positions. However, the higher cannot survive without the support of the lower; therefore, the humble should never be punished with violence for stepping outside of their position because that would make them unable to continue serving. Harmony takes precedence over victory. The loser must acknowledge the judge's decision, but the winner must acknowledge the equality of its opponent. This was an important lesson necessary to ensure a functioning society.

The Use of Ethos and Pathos

The sophists were known for the use of both ethos and pathos. All the disputations follow a pattern of establishing the speakers' ethos while attempting to discredit that of their rival. These appeals to ethos are compounded by their use of emotionally charged words to simultaneously appeal to their audience's emotions.

Use of Ethos

The use of the ethical appeal provides much of the structure of this disputation:

- Hoe attempts to destroy Plow's ethos—Plow doesn't serve the common good

- Plow attempts to rebuild its ethos—it serves the great which is better

- Plow attempts to destroy Hoe's ethos—it serves only the lowly, which is base

- Hoe rebuts Plow's criticism—Plow is dependent on Hoe. Hoe prepares the way for plow, who can't do its work without hoe. Not only does Plow require a team to keep it running, but it also only works four months to Hoe's twelve. Hoe argues that it is the basis of the city's economy. What could be more important than allowing the workers to feed their families so that they can serve their king?

Use of Pathos

Pathos, the appeal to the emotions, often relies on the use of specific sensory detail. This language engages the audience's imagination to encourage them to feel anger, fear, guilt, etc. In the disputation, both Hoe and Plow heavily embellish their claims with emotionally laden adjectives (see table 5.1).

Plow also uses many loaded adjectives, verbs, and phrases to appeal to its audience:

- great
- mighty
- faithful
- teeming
- adorning
- inspiring awe
- burrowing
- head always in the dust
- not fit for the hand of the noble

However, it is the importance of nomos that determines the outcome of the dispute. Next, I provide an analysis of *Hoe and Plow*.

Table 5.1. Hoe's Use of Emotional Language

Emotion	Text
Doubt	*Fragile* clay
Doubt	*Weakened* clay
Assurance	*Cool, well-built* dwelling
Hope	Put life into heart's again
Hope	Thirsty ones come back to life
Awe	The shepherd's hoe is surely set up as an *ornament*
Joy	"The Land watches me in joy!" (Vanstiphout 2003c, 578).
Gratitude	"Even the orphans, the widows and the destitute (50) Take their reed baskets And glean my scattered grains" (Vanstiphout 2003c, 579).
Disgust	"O Hoe, miserable hole digger, with your pathetic long tooth" (Vanstiphout 2003c, 579)

Source: Created by the author.

An Analysis of *Hoe and Plow*

According to Jarratt, oral poetry, such as the Homeric epics, "was structured on an echo pattern" . . . the parts of the poem are loosely connected and based on temporal, not causal links. The cultural instructions—nomoi and ethe—came to the audience in the form of a plurality of instances, not a generalized system" (Havelock 1963, 186). This is true in the *Hoe and the Plow*, in which the argument is presented as a series of examples that provide cultural instruction about the relative societal value of the poor versus the rich. In the disputation, the Hoe represents the commoner, while Plow represents the elite. The language used by each contestant makes this clear. The structure of both Hoe and Plow's arguments demonstrates the use of logos.

LOGICAL STRUCTURE

Hoe's argument follows a pattern.

What Plow Can Do

- Plow, you draw furrows—what is your furrowing to me?
- Plow, you make clods—what is your clod making to me? (Vanstiphout 2003c, 578)

What Plow Cannot Do

- You cannot dam up water when it escapes.
- You cannot heap up earth in the basket.
- You cannot press clay or make bricks.
- You cannot lay foundations or build a house.
- You cannot strengthen an old wall's base.
- You cannot put a roof on a man's house.
- Plow, you cannot straighten a street. (578)

A Repetition of What Plow Can Do

- Plow, you draw furrows—what is your furrowing to me?
- Plow, you make clods—what is your clod making to me? (578)

By surrounding the much longer list of what Plow cannot do, with the repetition of what it can do, Hoe effectively proves that Plow can only do *two* things. Further, it effectively minimizes the importance of the only two things Plow can do with the phrase, "What is your . . . . to me?"

Hoe is opening the argument by implying that the many (and more important) tasks that Plow cannot do, Hoe can. Or, Plow, you may be an expensive, specialized tool, but I, the common Hoe, have greater value because I do more with less. Clearly, the audience was meant to understand the same to be true of themselves. The farmer may work in the dirt, and the shepherd may work with animals, still kings are dependent upon them. Without them, there would be no kings.

Next Plow makes its first and only argument. It is interesting to note that while Hoe exclusively used second person "you" in its argument, Plow begins every example with "I," a stylistic choice that reinforces the cultural lesson that it is wrong to be self-centered and too sure of your own superiority. Plow's argument follows a different pattern.

Three "I" Statements

- I am Plow.
- I was fashioned by the great powers, assembled by noble hands.
- I am the faithful farmer of Mankind! (Vanstiphout 2003c, 578)

Ten Examples of What Others Do for Plow

- Even the King slaughters cattle for me,
- adding sheep!
- He pours out libation for me,
- And offers the collected liquids!

- Drums and tympans sound!
- The king himself takes hold of my handle-bars;
- My oxen he harnesses to the yoke;
- Great noblemen walk at my side;
- The nations gaze at me in admiration,
- The land watches me in joy. (578)

Five Examples of How Others Describe or Respond to Plow's Work

- The furrow I draw is . . . an adornment.
- The teeming herds of Shakan kneel down before my ripened grain.
- The shepherd's churn is filled to the brim.
- With my sheeves . . . the sheep of Dumizi are sated,
- Stacks adorning the plains . . . inspire awe. (578–79)

Five Examples of What Plow Does or Provides for Others

- Stacks and mounds I pile up for Enlil;
- Dark emmer I amass for him.
- I fill the storehouses of Mankind;
- Even the orphans, the widows and the destitute take their reed baskets and glean my scattered grains.
- My straw, piled up in the fields/ People even come to collect that. (579)

Four Disparaging Descriptions of What Hoe Does Beginning With "O Hoe"

- O Hoe, miserable hole-digger, with your/pathetic long tooth.
- O Hoe, always burrowing in the mud,

- O Hoe, whose head is always in the dust,
- O Hoe-and-brickmold, you spend your days in/mud, nobody ever cleans you! (579)

Three Commands

- Dig holes!
- Dig crevices!
- navel-man dig! (579)

One Description of Hoe

- Hoe, you of the poor man's hand, you are/not fit for the hand of the noble! (579)

Two Threats Against Hoe

- The slave's hand is adorned with your head!
- And you dare to insult me?
- You dare to compare yourself to me? (579)

One Final Brag

- When I go out to the plains, every eye is full of/admiration (578–79)

Plow clearly bases its entire argument of its superiority on its greatness:

- Even the great (gods and kings) serve Plow.
- Everyone admires its work.
- Plow does serve others, but Plow intentionally lists who it serves by their level of greatness.
- Plow disparages Hoe for getting dirty as it works. Clearly dirt is a sign of dishonor.

- Plow *orders* Hoe to work in the dirt because it believes it has the authority to do so.

    ◊ I do not believe the use of imperative voice is an accident here,

- Plow concludes by shaming Hoe for serving slaves and the poor and reaches the (illogical) conclusion that it is superior because it is *served* by gods and kings. One of the reasons Plow loses the argument is because the message being taught is that it is greater to serve rather than be served—a logical conclusion. What would happen to a society of people who all expected to be served? Even the king and queen serve the people.

The remainder of the argument is made up of Hoe demolishing (a fitting job for a hoe) Plow's accusations. Hoe makes these major points:

- Because of the humble work it does "is wealth spread everywhere."
- Hoe's work must come before Plow's making Hoe necessary to Plow's purpose.
- Plow's work is little, "though [its] ways are great" (579). Plow works four months to Hoe's twelve and its delusions of greatness are more than its amount of work.

This disputation is also a very interesting example of how nomoi were taught.

Use of Nomos

The economy of Mesopotamia was based largely on the growing of crops. As agrarian technology caused the population to grow, the economy became more complex. However, even rulers understood that the economy's success was based on the work of the commoner—not the contributions of the few and the great. This disputation effectively teaches the lesson that even the lowliest laborer should take pride in their work because it is essential to the well-being of all. And more importantly, as the god Enlil states at the disputation's conclusion, Hoe is great because it is humble and willing to work. Hoe is "Like a maid-servant, always ready, you will fulfill your task" (Vanstiphout 2003c, 580–81). Yet, Plow also serves a role. Plow may follow Hoe, but each is dependent on the other, as is every member of a functioning society. This message, a concise lesson in social justice, is just as important today as it was in ancient Mesopotamia—perhaps more so.

Conclusion

Many scholars choose to credit the Greeks with the invention of rhetoric because they were the first to conceptualize it. Certainly, this is true, but a failure to explicitly name what you do should not exempt you from credit for doing it. The ancient Mesopotamians did not leave us with the sorts of techne developed by the Greeks. However, the content of the clay tablets they did leave us clearly show that they taught rhetoric and practiced rhetoric in their written documents. Would we have more than seventy scribal copies of the same disputation if copying it wasn't a core activity in the scribal schools? Would the existing disputations follow a consistent pattern that marks them as a genre if scribes weren't taught that pattern? Is it coincidence that each disputation teaches important lessons that uphold the social contract:

- Don't commit violence
- Don't oppress the poor or the weak
- The great must rely on the humble

The words "mythos," "logos," and "nomos" have not been found in any Cuneiform writings. It is also difficult to know if, and to what extent, the Greeks might have known of the writings of the Sumerians. Whatever the answer to this question, based on the knowledge we have today, we should not fail to give the Sumerians credit for what they did—create the first rhetorical writings in the world.

Chapter 6

Myth, Magic, and Medicine
Medical Writing in Ancient Mesopotamia

> What is to be put on his forehead; hairs of a dog. And the head of the PQQ and its shoot he is to drink (mixed) together with fresh olive oil.
>
> —Pardee, *Ilu on a Toot*

You and some friends drop by the local beer merchant on your way home from work. You each order a flask of beer and pull out your personal straws to consume it with. One beer becomes two, and before you know it, you're stumbling home in the dark. Tomorrow will not be pretty. Good thing your dog has lots of hair.

Most of us are familiar with the expression, "hair of the dog," as a cure for a hangover: you need a drink to cure the effects of too much drink. But did you know that expression comes from a Sumerian poem called *Ilu on a Toot*, which literally tells the reader to put dog's hair on their forehead to cure a hangover? Dennis Pardee describes this poem as para-mythological text: one that features mythological characters but offers practical advice. The tablet on which this poem was written is divided into two parts. The first part tells the story of how the god Ilu got drunk at his local "drinking club" and had his hangover treated by the goddesses Anatu and Attartu. Pardee believes that this is more than a funny drinking story: "It appears as plausible to see here a serious attempt, even a scientific one, according to the science of the time, at dealing with the aftereffects of an evening spent on the benches of the *mrzh* [drinking establishment]" (Pardee 2003, 302–4). This is the type of medical advice we expect to find in a modern magazine or maybe on social media. This story might have been passed around in a similar fashion when it was written. And the recipe for the treatment was probably effective, even if it was a placebo effect. But this sort of anecdotal medicine was not the only type practiced despite what people believed decades later.

The Greek historian Herodotus wrote that the Mesopotamians "bring out all their sick into the streets, for they have no regular doctors. People that come along offer the sick man advice . . . no one is allowed to pass by a sick person without asking what ails him" (Kriwaczek 2012, 198). Luckily, Herodotus's description of Mesopotamian medical practices was grossly exaggerated. The Mesopotamians recognized two types of doctor: the *asu* and the *asipu*. And the Code of Hammurabi stated that a patient's ability to pay determined the fee a doctor charged (198), which might have made health care more accessible, though it did not mean that a doctor was obligated to treat someone. Most significantly, while their medical practice did blur the line between myth, magic, and medicine, that does not mean that medicine wasn't practiced by trained practitioners using recognized medical texts.

The earliest medical incantations come from Sumer—many of these were used to cure sickness of the mind, body, and community. While these incantations remained in flux for many centuries, by the first millennium these incantations had been "canonized" for use by the *asu* and *asipu*—the doctor and the diviner. These medical texts were written for both students of medicine and practicing doctors, and both their organization and content demonstrate a sophisticated understanding of document design, audience awareness, and the foundations of medical practice.

Medical Practitioners

The *asu* is the equivalent of today's general practitioner (Stol 2016). They were familiar with medicinal plants: where they grew, how to prepare them, and how to use them. Texts written for *asus* would have focused on that sort of information. If someone had a simple illness such as a headache or cold, they would have gone to the *asu* first, much as we would go to our general practitioner for medication and advice when we have a common illness such as the flu.

The *asipu* also diagnosed and treated illnesses. However, unlike the *asu*, the *asipu* also used divination and ritual to treat illnesses that had "magical" causes (Stol 2016). The Mesopotamians attributed many illnesses to the gods, ghosts, and other supernatural causes. Both the *asu* and *asipu* used a recognized series of medical texts as they trained to be doctors and later as they practiced.

Medical Texts

> The plant that resembles "sunflower" (but) whose seed is like (that of) *åigguåtu* is called *imœur-eåra*. **It is good for a persistent sore**. You grind (it and) pour it on the sore.
>
> —Scurlock 2014, 283

In early Mesopotamian history, medical lore, while shared, taught, and written down, had not been codified. It seems safe to assume medical lore was primarily an oral legacy passed from practitioner to student. As medical knowledge developed, however, a system had to be created for recording the ever-growing body of knowledge, but the Mesopotamians did not write information without purpose. They only recorded what could not be remembered. Therefore, we can only assume that simple procedures and processes were not included in the medical canon. Three series were produced:

- The Diagnostic and Prognostic Series
- The Vademecum—Pharmacist's Companion
- The UGU Catalog—The Therapeutic Series

THE DIAGNOSTIC AND PROGNOSTIC SERIES

King Adad-apla-iddina's consulting physician, Esagil-kin-apli (1068–1047 BCE) organized all the existing medical texts concerning diagnosis and treatment into a handbook made up of forty tablets: the number forty was associated with Ea, the god of healing. The handbook was divided into six sections:

1. Prognostication—ominous events observed by the *asipu* on their way to see a patient.

2. Anatomy—from the top of the head to the bottom of the foot. The fact that some tablets in the series focus on a specific type of illness (diseases of the eyes) suggests that doctors might have specialized in the types of treatments they offered as today's doctors do. "If his head, his body (and) the bulb of his nose continually give a jabbing pain [his] lips [make babbling noises(?)], among his people someone who died of thirst [afflicts him]" (Scurlock 2014, 22)

3. Time factors—how the length of time a person has been ill affects their prognosis and treatment. Also, phases of an illness, and the god, goddess, or demon causing the illness. "If he cries out with all his force, he is sick for one or two days and his crying gets softer, he will die." (192)

4. Neurology—conditions of the mind and body. "If when (a confusional state) comes over him, depression afflicts him (and) spittle flows from him mouth, a vow made by his father afflicts him; he will die." (Scurlock 2014, 201)

5. Infectious diseases—primarily focused on contagious diseases. "If the nature of the [sore] is that his body is full of what look like fish scales and it has a predictable course, it is called *risiktu*." (236)

6. Obstetrics/gynecology/Infants—fertility issues, sexually transmitted diseases, how to predict the sex of a child, and the treatment of a child's illness. "If an infant feeds from the breast and is not sated but drools a lot, his abdomen has been severely injured." (264)

The anatomical organization of the second subseries (anatomical principles) is attributed to EsagilkIn-apli, since neither the surviving forerunners nor the outlying diagnostic texts are organized in this fashion (Scurlock 2014, 8). Mesopotamian doctors relied almost exclusively on the use of plants for treating disease. *The Pharmacology* provided instructions on the properties of different plants.

THE PHARMACOLOGY

The Pharmacology lists plants to be used, followed by an explanation of which problems can be treated by those plants: "*Kamantu* seed is a plant to have seed. It is to be ground (and) given to drink (mixed) with first quality beer" (Scurlock 2014, 277). When treating patients with a commonly recognized illness, an allergy, a cold, or a boil, doctors started by offering a standard treatment of herbs like those that could be found in *The Pharmacology*. These treatments were simple and cost effective so they could give them to patients who might not be able to afford more specialized treatment.

THE THERAPEUTIC SERIES

The Therapeutic Series was a reference work meant to accompany the diagnostic series and help the doctor quickly scan symptoms (Scurlock 2014). Because it was a companion piece, it uses the same head-to-toe organization: "[If a person's] crown of head is continually hot, you mix [together] *gassu*-gypsum, *indar*-type(?) *uhhulu qarnanu, kibritu*-sulphur, bone, *uhhulu qarnanu,* rancid oil, fish oil. You fumigate his head (with it) over *asagu*-thorn coals" (322). Each option was listed sequentially, so a doctor could quickly scan for the closest match to their patient's symptoms, or identify a new treatment if a previous one had failed.

THE MEDICAL PROCESS

Basically, a medical student or doctor would have followed this process:

1. Become familiar with the *Diagnostic and Prognostic Series*
2. Become familiar with the content of *The Pharmacology*
3. Become familiar with *The Therapeutic Series*

For example, when treating a patient, a doctor would consult the *Diagnostic and Prognostic Series*. The doctor would ask the patient to describe their symptoms and then look for them in the handbook: "If he has a vise-like headache and his bowels are loose, 'hand' of Kubu" (34). Once they had diagnosed the illness, the doctor would have consulted *The Therapeutic Series* to discover a treatment: "[If a person'] crown of the head is continually hot, you mix [together] *gassu*-gypsum, indar-type(?) *uhhulu qarnanu, kibritu*-sulphur, bone, *uhhulu qarnanu,* rancid oil and fish oil. You fumigate his head (with it) over *asagu*-thorn coals" (322). As they prepared to treat the patient, the doctor might have consulted *The Pharmacology*: "*Purupuhu* is a plant for 'sick intestines"/being sick with *sahhu*. It is to be ground and given to drink (mixed) with first quality beer" (278). I have made this process sound simple. It's not. These tablets contain thousands of entries, and Cuneiform tablets were crammed with text with only the occasional horizontal line to break up sections. Navigating these materials would have been very difficult. And consider the difficulty of carrying around pounds and pounds of clay tablets as you made your rounds. However, even with the limitations naturally caused by the method of delivery the writers of the medical texts had to use (i.e., clay tablets), they developed and used methods to make the information as accessible as possible for their readers:

- horizontal lines and charts
- useful repetition
- sequential references
- side-by-side comparisons
- cross-references

Use of Horizontal Lines and Charts

As modern writers, we give little thought to the tools we use. We don't have to produce the electricity that powers our computers, and a quick trip to almost any store provides us with pencils, pens, paper, and whatever else we need. This glut of materials makes it easy for us to sacrifice space on a page, whether physical or virtual, to ensure that we use a pleasing amount of white space. But imagine being a scribe in ancient Mesopotamia.

Before you could write you had to shape clay into a tablet of the correct size, shape, and thickness. Before you could do that, you, or someone else, had to dig that clay out of the ground and rinse it to remove impurities. Once you made your tablet, you had to craft a stylus from a reed to use as a tool for carving Cuneiform figures into the wet surface of the tablet you had formed. Reeds had to be harvested from wetlands. The preparation for writing was just as labor intensive as the physical act of carving on wet clay, smoothing the clay to correct errors, drying the tablet, and then storing it. And if you dropped it, and it broke? You started the long process over.

When the act of writing is labor intensive and supplies are difficult to obtain, every inch of available space has to be used. So how did the ancient scribes attempt to bring order to the content of their documents without sacrificing room for more words? They used horizontal and vertical lines:

Figure 6.1. Medical text. *Source*: Penn Museum.

The pharmacology series used charts to categorize plants by

- Name
- Usage
- Method of preparation (Scurlock 2014, 273)

Table 6.1 is a transliteration of several lines from *The Pharmacology*. You can see that lines 24–29 are bracketed by horizontal lines at the top and the bottom. The line after 29 separates the last treatment for constriction of the urethra from the first treatment (30) for the gall bladder.

You can also see that there is a column for each category:

Table 6.1. Use of Lines and Columns in Medical Text.

	Name of Plant	"Ditto"	Use and Application
24.	ÚŠE.KAK GIŠDÌH	Ú KI.MIN	SÚD ina KAŠ.SAG NAG
25.	Úimhur-lim	Ú KI.MIN	SÚD ina GEŠTIN NAG
26.	Úa-su-pi-ru SIG	Ú KI.MIN	SÚD ina KAŠ.SAG NAG
27.	Úal-la-an-ka-niš	Ú KI.MIN	SÚD ina KAŠ.SAG NAG
28.	ÚSUMSAR	Ú KI.MIN	SÚD ina l+GIŠ u KAŠ. SAG NAG
29.	Úha-šá-a-nu	Ú KI.MIN	SÚD ina l+GIŠ u KAŠ. SAG NAG

Source: Scurlock 2014, 275.

The translation below demonstrates that each entry begins with the name of the plant, its use, and its application.

> 24. *Baltu*-thorn shoot is a plant ditto (for constriction of the urethra). It is to be ground (and) given to drink (mixed) with first quality beer.
>
> 25. *Imhur-lim* is a plant ditto (for constriction of the urethra). It is to be ground (and) given to drink (mixed) with wine.
>
> 26. Fresh *azupīru* is a plant ditto (for constriction of the urethra). It is to be ground (and) given to drink (mixed) with first quality beer.
>
> 27. *Allānkaniš*-oak is a plant ditto (for constriction of the urethra). It is to be ground (and) given to drink (mixed) with first quality beer.
>
> 28. *Šūmu* garlic is a plant ditto (for constriction of the urethra). It is to be ground (and) given to drink (mixed) with oil and first quality beer.
>
> 29. *Hašānu* is a plant ditto (for constriction of the urethra). It is to be ground (and) given to drink (mixed) with oil and first quality beer. (Scurlock 2014, 278)

Myth, Magic, and Medicine | 141

The writers of ancient medical texts also used several forms of repetition to help readers find information quickly, a technique we use today.

Useful Repetition

Imagine you are a doctor treating a critically ill patient. Time is of the essence, yet to find the answers you need, you must rely on the information etched on a thick stack of clay tablets. There is no table of contents or index. Each tablet looks much like the next. How do you find what you need in time? The consistent use of repetition would have helped. These descriptions are from *The Pharmacology*.

> 30. *Œiburu*-aloe is a plant for the gall bladder. It is to be ground and given to drink (mixed) with first quality beer.
>
> 31. *Ittu* is a plant for the gall bladder. It is to be ground and given to drink (mixed) either with first quality beer or wine.
>
> 32. *Merzinu* is a plant for the gall bladder. Ditto (It is to be ground and given to drink mixed either with first quality beer or wine).
>
> 33. *Kam,n åadê*-fungus[18] is a plant for the gall bladder. Ditto (It is to be ground and given to drink mixed either with first quality beer or wine).
>
> 34. *Tullal, bĭnu*-tamarisk leaves (and) snake skin(!) are plants for the gall bladder. Ditto (It is to be ground and given to drink mixed either with first quality beer or wine). (Scurlock 2014, 278)

Each of these five entries begins with the same opening line:

[name of plant(s)] is/are for the gall bladder.

The grammatical structure of the sentence places the name of the plant as the subject followed by the *to be* verb and subject complement. The consistent use of this a=b formula makes searching the various options easier.

Because pieces of information are repeated many times throughout the medical guides, often sequentially, the writers also made use of the word "ditto" rather than repeating a particular string of words:

> 30. *Siburu*-aloe is a plant for the gall bladder. It is to be ground and given to drink (mixed) with first quality beer.
>
> 31. *Ittu* is a plant for the gall bladder, It is to be ground and to drink (mixed) either with first quality beer or wine.
>
> 32. *Merzinu* is a plant for the gall bladder. Ditto.
>
> 33. *Kamun sade*-fungus is a plant for the gall bladder. Ditto.
>
> 34. *Tullal, binu*-tamarisk leaves (and) snake skin (!) are plants for the gall bladder. Ditto. (Scurlock 2014, 278)

In this example, in line one the text explicitly states: it is to be ground and given to drink (mixed) with first quality beer. In line two, this is changed to "with first quality beer or wine." However, in lines three through five, the text simply states, "ditto"—the text in parentheses was added by the editor and is not part of the original text. Useful repetition also exists in the lists of contents included at the beginning of each subseries within *The Therapeutic Series*:

> (obv. 12') [Total of six tablets of (the subseries called): "If a person's nose/mouth (feels) heavy," including] his nose/mouth (feeling) heavy, "sick lungs," *siqu* (colored sputum) and an infant sick with *su alu*-cough.
>
> (obv. 13'–14') : (The first tablet is called): "If a person is sick with *su alu*-cough and it turns into *kis libbi*. (The second tablet is called): "If a person's stomach is sick" (The third tablet is called): "If a person's upper abdomen (epigastrium) hurts him and when he belches, he continually produced bile, that person has "sick insides. (The fourth tablet is called): If *setu* 'gets' a person (and) he is sick with pulsating of the temples." (The fifth tablet is called): "If a person's inside is afflicted by fever." (The sixth tablet is called): "If a person's upper abdomen

(epigastrium) is puffed up (and) his hips (and) limbs continually hurt him." (Scurlock 2014, 300)

Each of the subseries within the larger series begins with a precis of this sort. Today, when we teach professional writing, it is common to tell students to include a roadmap in the introduction of the documents they write. The purpose of the roadmap is to tell the reader what major sections are included in the upcoming document, so they have context for the organization and know what to expect. We teach this practice because we know that roadmaps increase reader understanding and the ability to access the material they need.

The writers of these ancient texts used this technique for the same reason. *The Pharmacology Series* includes a huge amount of information, and its organization doesn't facilitate the matching of a treatment with a disease. *The Diagnostic Series* was written as a reference work meant to accompany *The Pharmacology* and make it easier for doctors to match treatments with diseases. These "roadmaps" achieved that goal.

It's hard to manage how difficult it would have been to search through stacks of large clay tablets to find the information you need when you had no table of contents, index, headings, or even pages. But the writers of these medical texts not only imagined these issues but also attempted to solve them by using techniques we still use today as professional writers.

Sequential References

Many different plants were used to treat the same illness, which makes sense. If the first plant doesn't work, you try the next. Or, if a plant is not available, another option exists. Doctors today also have multiple medications for the same illness for these very reasons. However, today's doctors have electronic databases they use to find and compare the available drugs. The writers of *The Pharmacology Series* also anticipated doctors' need to know all the available drug options before they chose one for treatment, so they used sequential organization. For example, nine different plants are listed as treatments for "constriction of the urethra." Because these nine options are listed one after the other and separated from other text on the tablet with horizontal lines, a doctor could easily compare each option before treating a patient. This organizational method is also used in *The Therapeutic Series*, which lists every possible treatment for the same illness in one consecutive list:

> (5–17) If a person's eyes become dimmed, you grind 1.2 shekel "white plant," 2 shekels *rikibti arkabi*, 15 grains of alum, 15 grains of *emesallim*-salt, (and) 1.8 shekel of *kanaktu* "fat" together in *gunnu* oil and daub it on his eyes.
>
> (18–19) If a person's eyes become dimmed and contain tears, you grind "white plant" (and) wild honey in oil (and) daub on.
>
> (20–21) If a person's eyes become dimmed, you grind myrrh, "white plant," (and) *rikibti arkabi* ghee (and) daub it on his eyes. (Scurlock 2014, 364)

Again, this is not random organization. The writers intentionally considered their audience and used a method of organization that would make the information more accessible.

Medical Commentary

Sometimes even these devices were not enough. Several medical commentaries exist. These commentaries are the notes and explanations of different doctors as they attempted to understand and apply what they were learning. Many of the comments are definitions of terms, or helpful hints for recognizing different medicinal plants:

> Eristu-fat means marrow from the short-bone (of sheep or goats). (Scurlock 2014, 339)
>
> "Hand" of a goddess—(when) he continually has a crushing sensation in the chest and continually forget his words, it is "hand" of a goddess. (343)
>
> *Kurkanu*-tumeric resembles a shaved armpit. (345)

A colophon found on one of the commentaries is actually a commentary on a commentary. Could this be the first use of meta-text? "Commentary (based on) oral tradition and questions from the mouth of an expert in it" (356).

Medicine and Social Justice

Despite what Herodotus believed, access to medical treatment was not guaranteed to all because doctors charged a fee for treatment. However, the Code of Hammurabi did include "laws [that] specified the fees to be paid to physicians, depending on the status—and therefore resources—of the patient" (Kriwaczek 2012, 198). Unfortunately, we have no way of knowing if this law was enforced, though the fact that the law was even written tells us a great deal about the society's values. The system of payment was simple: the more complex the illness, the more expensive the treatment. The simpler the diagnosis, and the fewer and more abundant the plants used for treatment, the less expensive the service (Scurlock 2014, 273). But the simple fact that cost was tied to the complexity of the treatment suggests that many patients would have had to settle for treatments that were ineffective because they were cheaper.

To make a modern parallel, the level of treatment a person receives is often dictated by the type of care offered by their country: insurance, socialized, fee based, etc. In the United States, a person's medical coverage is determined by whether they have insurance and what that insurance covers. Or, if they live in a country that offers socialized medicine, the care they receive is based on what the system allows, while their ability to subsidize that care depends on their ability to pay. And often, patients must settle for less effective medications because their insurance will not pay for the more expensive option.

In addition to asking who had access to treatment, we need to consider who had access to medical training.

Could Women Be Doctors?

We do have evidence that women did work as *asu* (doctors). In the Ur III period, a tablet lists a doctor with a female name. The names of two women doctors were also included on tablets found at the court of Mari, and an Old Babylonian list references a woman doctor and a woman "midwife" (Stol 2016, 371). The ritual below used to induce labor specifically names a midwife: "Recitation: Run to me like a gazelle; flee to me like a little snake. I, Asalluhi, midwife; I will receive you" (Scurlock 2014, 129).

Sadly, the presence of a handful of women doctors and midwives does not suggest that women had equal access to medical training. It is a typical feature of Mesopotamian magic and medicine that the two professionals participating in the textual tradition—exorcist and physician—are both male; whereas, female healers are restricted to the divine sphere (the healing goddess Gula), the lower rungs of society, and the stereotypes associated with the agents of harmful magic—witches. But this pattern is by no means universal in the ancient Near East. In Anatolia, male and female physicians are attested, and the main bodies of therapeutic rituals are assigned to various groups of ritual experts, the most prominent among which are the "old woman" and the "diviner" (Schwemer 2018, 28).

Conclusion

When you have a stuffy nose and a cough, you probably go to your local pharmacy to pick up some over-the-counter medicine. You might even consult the pharmacist if you aren't sure what to buy. The pharmacist, who is a trained professional, would consult a database to find you an effective treatment, and it would not be at all surprising to find honey in the cough medicine or eucalyptus in the decongestive. However, when you are taking those medications, you might not have realized that another person, thousands of years ago in Mesopotamia, likely took the same ingredients for the same purpose as prescribed by the *asu*.

And when you begin to run a fever? You might schedule an appointment with your doctor, who will rely on years of training to prescribe medication—probably a fever medication and an antibiotic. In 2000 BCE, the *asipu* would have prescribed willow bark for the fever and echinacea as an antibiotic. Both willow bark and echinacea are still used by modern doctors. The *asipu* would have charged you significantly more for their specialized treatment. Both the medicine you bought over the counter and the medicine you received from the doctor would have included instructions for use. Some things never change, do they?

Chapter 7

Writing as Social Justice

> It should be mentioned, moreover, that the written word was also in the service of power. It was used largely by kings and their administrations to control the population.
>
> —Charpin, *Writing, Law, and Kinship in Old Babylonian Mesopotamia*

You are a young slave boy of eight. Yesterday, your master purchased a camel from your neighbor. After your master and the other man sealed the deal, they went to the local tavern to drink beer. Because it was very late before they returned home, the camel remained in your neighbor's yard. This morning, your master told you to go and get the camel and bring it home. When you get to your neighbor's home, you see the camel tied to a fence. You untie the camel's rope and begin to lead him home. When you are halfway to your master's house, you are stopped by two men you don't recognize. "Thief!" one shouts at you, "you are stealing our brother's camel." "No," you reply, "my master bought this camel." "Then where is your receipt? Where is your seal? We will take you to the big man of the village, and he will punish you." You shake with fear. If you run to your master, he will punish you, but if you stay with these men, they will do the same. What do you do?

The members of the TPC community do not unanimously agree that social justice should be either a concern or a subject for research because some believe technical communication's objectivism makes it immune to such transgressions: "Because technical communication often appears to be removed (sometimes twice removed) from the atrocities of domination, the field can maintain (and has maintained) its distance from the violence, oppression, and injustices it perpetuates" (Walton, Moore, and Jones 2019, 17–18). Clearly, not everyone agrees that TPC deserves protected status,[1] though as Walton, Moore, and Jones remind us, our field "historically has done little to articulate oppression as a central concern" (18). However, because TPC research historically centers on communication as the nexus between product and consumer, any attempt to pretend social justice is not a by-product of TPC is in fact an act of oppression because products and services have an oppressive effect that "can be easily ignored and dismissed by designers" (18) and documenters of these products and services. Let's look at an example.

The internet, as both a product and a service, suppresses the opportunities of those who don't have it. Sixty-three percent of the world's population has access to the internet. Affluent people in highly populated areas of developing countries claim most of the access. A technological gender gap also exists. Globally, men account for 18 percent more social media use than women, and in Southern Asia, men are two and a half times more likely to use it. When women are denied access to the internet and digital technologies, they lose the chance to start businesses, sell products, find jobs, get an education, and generally participate in a public life (Kemp 2022). Unless it can separate itself from the use of the internet (and it can't), TPC cannot claim that it does not participate in the suppression of those who are not being allowed to participate. TPC does not have the right to pretend it does not directly influence the power struggle currently being waged between those who would control internet access and those they've attempted to deny. As TPC practitioners and scholars, we are complicit. We have always been complicit.

If we take a historic view, it is apparent that the people who first invented writing also used it for oppression, which means that writing was the catalyst for social justice because "Oppression makes social justice necessary" (Walton, Moore, and Jones 2019, 16).

In ancient Mesopotamia, "The written word was also in the service of power. It was used largely by kings and their administrations to control the population" (Charpin 2010b, 3). The word "control" describes a form of oppression, and consequently, unless they could also harness the power of the written word, those people who were not royalty or part of their administration were easy targets for oppression. The conundrum is the simple fact that the very written language that suppressed people was also their only tool for seeking social justice: the same game of double jeopardy we continue to play today.

The "control" exerted by writing often came in the form of written laws—and a large portion of the texts that have been recovered are legal in nature (Charpin 2010b). However, in this age when humans were still on the cusp of the transition from orality to literacy, not everyone was literate. More people than we had at first thought could likely read and write: scribes, priests, military, merchants, and others. And still more could read but not write. But what of those who could do neither and who could not hire a scribe or go to a friend who had the skills they lacked? What use is a law that can't be read? What recourse for defense does someone who can't read or write have? Look at this law from the Old Assyrian period: "If she [the first wife] has not produced a descendant for him within two years, she shall herself buy a slave girl and as soon as she [the slave girl] later produces a child for him, she [the first wife] may sell her [the slave] to whomsoever she wishes" (Stol 2016, 169). The slave girl is clearly being raped, forced into childbirth against her will, and then separated from her child and sold—all without her consent. Yet, because a written law makes these actions possible, the young woman has no recourse. And if the young woman couldn't read, she wouldn't even be able to understand the law that bound her.

What if she ran away? Slaves were dressed so that they could be identified (Charpin 2010b): "The agents of the king have seized him (my young man) in the town of Appasum and he is being detained in the house of Nurum-lisi. But this man wears neither the fetters (of a slave) nor the hairdo of a slave" (Oppenheim 1967, 82). In this example, the dress code for slaves saved the young man, but the situation could easily be reversed.

Writing was also used in service of social justice. When two warring kings agreed on peace terms, they exchanged tablets on which they wrote their conditions for peace. If either party was suspected of breaking the peace terms, the tablets were used as proof either for or against the infraction. And Hammurabi regularly issued written edicts that offered amnesty for various crimes. He did this to placate Shamash, the god of justice (Charpin 2010b).

However, a careful consideration of both the tools they developed and the words they wrote quickly shows that writing was first, and in many ways most importantly, a tool of oppression (Young 2003). Young (2003) divided oppression into five categories: marginalization, cultural imperialism, powerlessness, violence, and exploitation. Textual evidence demonstrates that each of these forms of oppression occurred in ancient Mesopotamia and can be tied directly to someone's ability to engage in professional communication.

Beginning with the development and use of cylinder seals, to the evolution of Cuneiform itself, writing in all its evolutionary forms was used to both empower and enslave: "There is one fact which can be established; the only phenomena which, always and in all parts of the world, seems to be linked with the appearance of writing . . . is the establishment of hierarchical societies, consisting of masters and slaves, and where one part of the population is made to work for the other part" (Charbonnier 1961, 29–30). However, before we look at specific instances of oppression, we must understand how the culture itself understood oppression and social justice.

Oppression and Justice

According to Paul Kriwaczek, the urban revolution—the intentional shift from hunting and gathering to village then city life—was an ideological choice that required those who made this choice to give up "their autonomy, their freedom, and their very identity as self-reliant and independent actors" (Babylon 2012, 448). Artifacts alone can never fully reveal what this ideology might have been; however, archaeological records suggest that the people who built the first city, Eridu, valued progress simply for the sake of progress. The first building in Eridu was a small temple. The citizens destroyed and replaced the original structure eleven times over the next millennium (about every ninety years), never reproducing but always enlarging and improving. The last building was a temple the size of a monument. According to Kriwaczek: "The Eridu temple was the symbol of a community who believed in—perhaps one might even say invented—the ideology of progress: the belief that it was both possible and desirable to continually improve on what had gone before, that the future could and should be better—and bigger—than the past" (2012, 471). Progress gave birth to many other institutions including social classes, division of labor, organized religion, and writing, which developed to meet the recordkeeping needs of the increasingly complex society. Ultimately, the urban revolution, which spawned the city, and paved the way for scientific and creative exploration, also created the catalyst for organized oppression marked by changes in the system of leadership.

Before 4000 BCE: The Urban Revolution

The citizens of Eridu believed "that it is humanity's right, its mission, its destiny to transform and improve on nature and become her master" (Kriwaczek 2012, 20). For the next ten to fifteen thousand years, this group of people developed a centralized state that recognized a social hierarchy, division of labor based on that hierarchy, organized religion, monumental architecture, engineering, writing, art, education, mathematics, and law. Wheeled vehicles and sailing ships were invented, as well pottery and pottery kilns. At this point in its development, Mesopotamian culture was largely theocratic. Social oppression didn't just occur—it was believed to be ordained by the gods, who themselves suffered from oppression:

> 1–11. In those days, in the days when heaven and earth were created; in those nights, in the nights when heaven and earth were created; in those years, in the years when the fates were determined; when the *Anuna* gods were born; when the goddesses were taken in marriage; when the goddesses were distributed in heaven and earth; when the goddesses . . . . became pregnant and gave birth; when the gods were obliged (?) . . . . . . their food . . . . . . dining halls; the senior gods oversaw the work, while the minor gods were bearing the toil. The gods were digging the canals and piling the silt in *Harali*. The gods, crushing the clay, began complaining about this life. (Enki and Ninmah 2016)

According to this mythological composition, which explains the origins of the world and how it works, the gods themselves had a hierarchy in which the minor gods were oppressed by the senior gods. The minor gods blamed Enki, who created the senior gods. However, because Enki is sleeping and refuses to listen, his mother, who gave birth to the senior gods, wakes him, and demands: "Please apply the skill deriving from your wisdom and create a substitute (?) for the gods so that they can be freed from their toil!" (Enki and Ninmah 2016). Enki tells his mother, "The creature you planned will really come into existence. Impose on him the work of carrying baskets" (Enki and Ninmah 2016). He tells her that Ninmah the goddess of birth and the creator of humankind will assist her. However, before she begins, Ninmah and Enki get drunk. In their inebriated state, Ninmah tries to get the better of Enki by producing humans who are differently abled, believing Enki will reject them. Enki, however, proves their worth:

> Enki and Ninmah drank beer, their hearts became elated, and then Ninmah said to Enki: "Man's body can be either good or bad and whether I make a fate good or bad depends on my will."
>
> 56–61. Enki answered Ninmah: "I will counterbalance whatever fate—good or bad—you happen to decide." Ninmah took clay from the top of the abzu in her hand and she fashioned from it first a man who could not bend his outstretched weak hands. Enki looked at the man who cannot bend his outstretched weak hands, and decreed his fate: he appointed him as a servant of the king.
>
> 62–65. Second, she fashioned one who turned back (?) the light, a man with constantly opened eyes (?). Enki looked at the one who turned back (?) the light, the man with constantly opened eyes (?), and decreed his fate allotting to it the musical arts, making him as the chief . . . . . . in the king's presence.
>
> 66–68. {Third, she fashioned one with both feet broken, one with paralysed feet. Enki looked at the one with both feet broken, the one with paralysed feet and . . . . . . him for the work of . . . . . . and the silversmith and . . . .} {(1 ms. has instead:) She fashioned one, a third one, born as an idiot. Enki looked at this one, the one born as an idiot, and decreed his fate: he appointed him as a servant of the king.}
>
> 69–71. Fourth, she fashioned one who could not hold back his urine. Enki looked at the one who could not hold back his urine and bathed him in enchanted water and drove out the namtar demon from his body.
>
> 72–74. Fifth, she fashioned a woman who could not give birth. Enki looked at the woman who could not give birth, {and decreed her fate: he made (?) her belong to the queen's

household.} {(1 ms. has instead:) . . . . . . as a weaver, fashioned her to belong to the queen's household.}

75–78. Sixth, she fashioned one with neither penis nor vagina on its body. Enki looked at the one with neither penis nor vagina on its body and gave it the name 'Nibru eunuch (?),' and decreed as its fate to stand before the king.

79–82. {Ninmah threw the pinched-off clay from her hand on the ground and a great silence fell}{(1 ms. has instead:) Enki threw all (?) the clay to the ground and was greatly . . . . . .}. The great lord Enki said to Ninmah: "I have decreed the fates of your creatures and given them their daily bread. (Enki and Ninmah 2016)

This mythological explanation of the creation of humans presents an interesting puzzle. On one hand, we clearly see that the Mesopotamians believed and accepted that oppression of some by others was the norm. Further, how someone was forced to serve could be determined by their physical abilities, thus rendering them powerless. And powerlessness is one of five aspects of oppression (Young 2003). In this case, writing serves to oppress by stating that oppression is acceptable.

On the other hand, some justice is shown to the differently abled: rather than rejecting them as useless, which is what Ninmah clearly intends, Enki does the opposite and finds value in each of them. Unfortunately, that value comes with a high cost—they must serve others.

This narrative provides interesting insight into the society's understanding of social hierarchies. First, they did exist, and they were created by the gods who suffered under their own system of oppression. Second, the solution to oppression is to find someone else to oppress. Finally, because this belief was presented as part of the literary canon, it would have been commonly known by the literate public and probably even heard by those who couldn't read.

This same ideology is present in the disputation *Hoe and Plow*, which I discussed in an earlier chapter. The point of the dispute is to demonstrate that the humble are more important than the elite because the elite couldn't function without their service. However, this compliment is backhanded at best as the resulting message is the same—convince the humble that the appreciation of the elite is enough to justify their forced servitude.

Between 4000 and 3000 BCE: Temple Rule

The next great city to be built was Uruk. Many monuments were built in the city utilizing new inventions such as concrete. For thousands of years, the citizens of Uruk built, rebuilt, and replaced these great temples. However, no evidence suggests that the laborers were forced into servitude. According to Kriwaczek, "All the signs for this era point to a society with no overly great distinctions of wealth or power" (2012, 41). Instead, the citizens were organized around their worship of the Great Goddess of Uruk. As the city's infrastructure developed, people began to work for salaries, which led to merchants selling wares and ultimately to what is called "aesthetic of deprivation." As people's financial resources began to determine what they could own, and as need led to mass production of standard items, those people with limited resources had to buy what was utilitarian, not what was aesthetically pleasing. Imagine the difference between a paper plate and a piece of Limoges china. So even in a society that valued equality, oppression occurred.

The aesthetic of deprivation also affected the purchase of cylinder seals. Cylinder seals, like today's pens, came in many different qualities: crude "off the rack" seals to bespoke seals of hematite and lapis lazuli.

Some members of society were marginalized by the quality of their possessions. In fact, some scholars suggest that it was economic instability that led to the collapse of the city. Although the priests who led the city began with a vision of an egalitarian society in which everyone willingly served the city's goddess, as the number of hungry mouths to feed continued to grow, it is not hard to imagine that they began forcing those with the least power to work in the fields. Cylinder seals from this time show scenes of men being guarded by other men with spears. Were these men being forced to work for the city?

3000–2300 BCE: Big Men and Kings

After the fall of Uruk, competing rulers, though they still claimed a belief in a common culture, began to fight for territory. As a result, priests were replaced by warlords and kings, and the "relatively egalitarian society of religious rule [shattered] into classes of rich and poor, weak and strong" (Kriwaczek 2012, 78). Archaeologists have identified the remains of thirty-five towns and cities, with smaller villages between. Ownership of private property, particularly property that could be used as grazing land, became a source of wealth. Ownership of property was not restricted. Anyone, including women, who had the money, could purchase and own land. The wealthy could work or rent their land, sell what they produced on those lands, and sponsor workshops of artisans. In this period, the wealthy paid others to do their work for them, and the gap between the haves and have nots grew as a result. Leadership, however, fell on a new class of citizen—the big men.

The Big Men

Because of the constant warfare between competing cities and towns, each town needed to be able to defend itself. The rise of the "big men" was the result. These military leaders created and trained their own armies to protect their homes. They also used the written word to force men into military service: "Censuses played an essential role, allowing the authorities to register, locality by locality, men to be mobilized in case of conflict" (Charpin 2010b, 3).

However, even the big men needed a leader, the principal *Lugal*, or head big man. *Lugal*, which translates as king, has one profound difference: a *Lugal* was a mortal man, while a king was known to be descended from the gods. Gilgamesh was a *Lugal*, but the later Queen Puabi was a goddess. Evidence of this shift from *Lugal* to king or queen was partially based on the discovery of the royal graves of Ur. These graves included many grave goods of exquisite quality: gold jewelry, musical instruments, and weapons. These graves also included the bodies of the men, women, and animals, who were sacrificed—willingly or unwillingly—to accompany their ruler to the afterlife so they could continue to serve them. Here is a powerful example of oppression as the society devolved into one dominated by a series of god-kings who squeezed the citizenry and appropriated temple property (Kriwaczek 2012, 100)—acts that finally allowed an usurper, Urukagina, to seize control by staking his legitimacy on "his claim to [end] the exploitation of the common people by both palace and temple" (Kriwaczek 2012, 101).

The Rule of Urukagina

After taking power, Urukagina discovered rampant corruption: officials lining their own pockets, the royal family taking land from families, and the levying of heavy taxes:

> The inspector of the boatmen seized the boats. The cattle inspector seized the large cattle, seized the small cattle. The fisheries inspector seized the fisheries. When a citizen of Lagash brought a wool-bearing sheep to the palace for shearing, he had to pay five shekels if the wool was white. If a man divorced his wife, the ishakku got five shekels, and his vizier got one shekel. If a perfumer made an oil preparation, the ishakku got five shekels, the vizier got one shekel, and the palace steward got another shekel. (Kramer 1981, 48).

One anonymous citizen summed it up this way: "You can have a Lord, you can have a king, but the one to fear is the tax assessor."

After Urukagina took power, one citizen claimed that "there was no tax collector" (49). He also protected the rights of the ordinary: "The house of a lowly man was next to the house of a 'big man,' and the 'big man' said to him, 'I want to buy it from you.' If, when he (the big man) was about to buy it from him, the lowly man said, 'pay me as much as I think fair,' and then he (the big man) did not buy it, that 'big man' must not 'take it out' on the lowly" (49).

Although his reign lasted only ten years, his social reforms left their mark on history. Three clay cones and one oval plaque that are carved with the text of his reforms have been recovered. His code was followed by that of Ur-Nammu (112–2095 BC), who also attempted to provide protection for the weak by creating what he called "equity in the land": "The orphan did not fall a prey to the wealthy"; "the widow did not fall a prey to the powerful"; the man of one shekel did not fall a prey to the man of one mina (sixty shekels)" (Kramer 1981, 54). And while the laws listed are harsh, they show a movement away from and "eye for an eye" and toward a more humane approach to punishment:

> If
> (a man to a man
> With a . . . instrument)
> His . . .
> The foot has cut off,
> 10 silver shekels
> He shall pay.
> If
> a man to a man

with a weapon
his bones
of . . .
severed,
1 silver mina he shall pay.
If
a man to a man
with a geshpu-instrument
the nose (?) has cut off,
2/3 of a silver mina
he shall pay. (55)

An examination of the communication tools used by this society also shed light on both the oppressed and the oppressors.

Cylinder Seals

> Distributive paradigms place "stuff"—material goods, resources, benefits—at the center of justice considerations. This form of justice is concerned with people getting their fair share.
>
> —Walton, Moore, and Jones 2019, 14

In ancient Mesopotamia, the ability to fully participate in written communication required the ability to acquire the tools of literacy, beginning with cylinder seals. Because seals had to be either purchased or gifted, those who were unable to obtain seals suffered from distributive injustice, which places "material goods, resources, benefits—at the center of justice consideration" (Walton, Moore, and Jones 2019, 14). Because some people were denied access to cylinder seals, they couldn't get their fair share of the benefits associated with the practice of written communication.

In ancient Mesopotamia, one of the first and most powerful technological developments was the cylinder seal. Because cylinder seals were a person's only means of "signing" their name, possession of a cylinder seal was required for engaging in business or legal transactions. Consequently, cylinder seals as a means of communication became a form of oppression.

Anyone who could purchase a seal could own one—even a slave. Any person could also be gifted a seal. Seals, like all commodities, were of varying quality. The wealthy often purchased bespoke seals that depicted scenes from their lives that they considered important, or information that identified their profession. Other seals were mass produced in workshops and were carved with poor-quality generic scenes. So even though anyone could buy (or be given) a seal, the quality of that seal still had social implications for the owner, who would have been judged by the quality of their personal seal. Seals were used to literally "seal the deal." One or both of the people engaged in the transaction would roll their personal seal across the wet clay of the tablet, essentially "signing" their name. Occasionally, when someone who was required to "sign" did not have their seal with them, they would use someone else's seal, or even substitute the imprint of their fingernail (Charpin 2010b). However, in such cases an inscription would be added to the contract identifying the substitution that was made.

Cuneiform

Although scholars now believe that more people than originally thought were literate, we must remember that the term "literate" had a looser meaning than it does today. The highly educated, usually high-ranking government officials, priests, or scribes, could both read and write Cuneiform. However, if someone listened to a person read them a letter, that was considered a form of reading, just as dictating a letter was considered a form of writing. Anyone who had the money could hire a scribe to transcribe a letter for them. We know that even slaves did this:

> Tell my master: Your slave girl Dabitum sends the following message:
> What I have told you now has happened to me: For seven months this (unborn) child was in my body, but for a month now the child has been dead and nobody wants to take care of me. May it please my master (to do something) lest I die. Come visit me and let me see the face of my master! [Large gap] Why did no present from you arrive for me? And if I have to die, let me die after I have seen again the face of my master! (Oppenheim 1967, 85)

I find the last sentence of this message particularly heart wrenching. I doubt that a slave, especially one in this woman's condition, truly longed to see her master's face: Yet how else is she to persuade him to help her if not by pretending to have an emotion she likely didn't feel? She had to lie and debase herself to ask for the health care she should have been owed.

Scribes, Writing, and Access to Literacy

Most written documents were dictated to scribes, and the scribe's training and social position in relation to the person dictating the message would have affected the words they chose when recording the message. Some scribes were highly trained and had the social position to actively collaborate in the writing of a letter, while others were "poorly paid town scribe[s] who translate[d] the inarticulate complaints of the poor and uneducated into the stereotyped eloquence of a petition or begging letter" (65). However, regardless of the scribe's social standing, he or she had the power to work for or against the needs of the person who hired them, making the scribe a tool for obtaining, or preventing, social justice.

Literacy itself was known to be a useful tool for oppression. Babylonian law did not require contracts to be written, and many of the written contracts contained only the barest of details (Charpin 2010b). However, "The use of writing . . . made inroads because of the witnesses' morality" (49). The Babylonians understood that a clay tablet could be preserved, but they also learned that they could be destroyed, or edited, to take advantage of others.

During this period, the rulers periodically issued *misarum*: the canceling of debts. After such an edict was made, creditors were made to present their tablets to a commission, who broke them, canceling the debt of the person under contract. In year thirteen of Hammurabi, a woman creditor claimed to have lost her tablets. In order to prevent her from using this to avoid canceling her debtors' debts, the commission broke up a large clod of dirt as a symbolic act and sent each of her clients a tablet stating their debt had been canceled (Charpin 2010b).

Conclusion

Despite evidence to the contrary, many people still believe that written language can be objective and that its objectivity is somehow a vaccination against social injustice. Language is many things: powerful, persuasive, pernicious. But it is never objective.

I regularly teach a course on writing procedures and manuals. On the first day, I like to share this article with my students: "Poor Instructions Cause Loss of Finger":

> The patient in this case was a 2-year-old female. A physician assistant (PA) at a pediatric practice performed a finger stick test on the patient for anemia and lead poisoning. Following the test, the PA applied gauze and a bandage to the patient's finger, and the family returned home. Three days later, the patient's family brought the patient to an urgent care clinic because she was in pain. The child's finger was still bandaged, and it had turned necrotic, requiring a partial amputation.

What happened? The PA made an assumption about her audience:

> In this case, the absence of explicit verbal guidance and written follow-up instructions from the PA resulted in the parents leaving the child's bandage in place, which cut off blood supply to her finger and caused tissue death. The assumption that the parents would understand to remove the bandage—because the child would stop bleeding in a short amount of time, and the bandage could impede circulation—was a lapse in judgment by the PA.

The PA did not know the parents of that child. They did not know what level of familiarity they had with medical procedures—even one as simple as a finger prick. They did not know the parents' familiarity with either spoken or written English. Yet they acted as though they knew their audience well enough to assume they knew how to care for their child's finger. Because of their choice to assume rather than ask, a child lost a portion of her finger. Is that justice?

When we fail to understand that our intentions as we write mean nothing because every reader will bring their own meaning to those words, we fail to recognize our responsibility as writers. And this failure results in a lack of justice for all or part of our audience. We can choose to pretend we have no obligations as producers of TPC, but that choice is itself a form of injustice.

Notes

Chapter 2

1. See https://examples.yourdictionary.com/examples-of-conditional-conjunctions.html.

2. I am excluding the second clause from line 1 because it contains a word in parentheses: "(take) 2 seahs worth of cane, along with their *tubaqu*-roots (i.e., the whole cane)."

3. I find it interesting that Oppenheim uses the word "argument" without explanation as though it is a given that the Mesopotamians were writing argumentatively and developing "lines of argumentation used to present pleas" (1967, 65). However, this is the subject of another (and longer) project.

Chapter 3

1. The naditu were unmarried women, who lived separately from the rest of society and served a specific god or goddess. Families often provided a dowry so that a daughter might become a naditu.

2. Most scholars believe that after 331 BCE the Greek influence dominated speech and writing (Stol 1995). Richard Enos, however, argues that the Greek influence begins much earlier: by the latter half of the eighth century BCE, "the techniques of composing discourse exclusively for an oral medium were beginning to be replaced by developing scripts" (Enos 1993, 2).

Chapter 4

1. See the following. Gerd Brauer; Christiane Donahue; Andrea Scott; C. Bazerman, C. Early, J. Lundsford, S. Null, P. Rogers, A. Stansell; B. Horner,

S. NeCamp, and C. Donahue; J. Kearns, and B. Turner; S. Lahm; D. S. Martins; A. Scott; C. Thaiss, G. Brauer, P. Carlino, Ganobcsik-Williams, and A. Sinha; Lisa Emerson, Bruce MacKay, Keith Funnell, and Marion MacKay; Bruce Horner, and John Trimbur; Min-Zhan Lu; LuMing Mao; Paul Kei Matsuda; Jonathan Monroe; Mary Muchiri, Nshindi Mulamba, Greg Myers, and Deoscorous Ndoloi; Karen Ogulnik; Paul Stapleton; Marty Townsend; Xiaoye You.

2. See *College English* 78, no. 3 (January 2016), for a comprehensive discussion of the translingual writing movement and its impact on the teaching of writing.

3. The naditu were unmarried women who lived separately from the rest of society and served a specific god or goddess. Families often provided a dowry so that a daughter might become a naditu.

Chapter 5

1. Millman Parry, *The Making of Homeric Verse: The Collected Papers of Millman Parry*, ed. Adam Parry. Oxford: Clarendon, 1971; Albert B. Lord, *The Singer of Tales* (New York: Athenaeum, 1976).

2. Dissoi Logoi is the practice of arguing multiple sides of one question, and sometimes making the worse seem better or the better worse. For example, in *Hoe and Plow*, Plow makes the act of getting dirty worse, while Hoe makes it better than being clean and lazy.

Chapter 7

1. See Walton et al., 2019, 18.

Bibliography

Alster, Bendt. 1975. "Paradoxical Proverbs and Satire in Sumerian Literature." *Journal of Cuneiform Studies* 27, no. 4: 201–30.
———. 2005. *Wisdom of Ancient Sumer.* Bethesda, MD: CDL Press.
Babcock, Barbara A. 1984. "Arrange Me Into Disorder: Fragments and Reflections on Ritual Clowning." In *Rite, Drama, Festival, Spectacle: Rehearsals toward a Theory of Cultural Performance*, edited by John J. MacAloon. Philadelphia: Institute for the Study of Human Issues.
Bahrani, Zainab. 2011. *Women of Babylon: Gender and Representation in Mesopotamia.* London: Routledge.
Bayliss, Miranda. 1973. "The Cult of the Dead King in Assyria and Babylonia." *Iraq*, 115–25.
Bellos, D. 2011. *Is That a Fish in Your Ear?* New York: Farrar, Straus, and Giroux.
Besnier, Marie-Françoise. 2014. *The Geography of Knowledge Project.* University Park: University of Pennsylvania. Accessed July 2021. http://oracc.museum.upenn.edu/cams/gkab/P338688.
Bizzell, Patricia, and Bruce Herzberg. 1990. "General Introduction." In *The Rhetorical Tradition: Readings from Classical Times to the Present*, edited by Patricia Bizzell and Bruce Herzberg, 1–16. Boston: Bedford Books.
Boje, David M. 2001. *Narrative Methods for Organizational and Communication Research.* Thousand Oaks, CA: Sage.
Boquet, Elizabeth H. 2002. *Noise from the Writing Center.* Logan: Utah State University Press.
Bratta, Phil, and Malea Powell. 2016. "Introduction to the Special Issue: Entering the Cultural Rhetorics Conversations." *Enculturation: a Journal of Rhetoric, Writing and Culture* 21.
Brauer, Gerd. 2002. "Drawing Connections Across Education: The Frieberg Writing Center Model." *Language and Learning Across the Disciplines* 5, no. 3: 61–80.
British Museum. n.d. *Cylinder Seal.* London: British Museum.
———. 2019. "British Museum Collection Online." *British Museum Online Collection.* August 20. https://britishmuseum.org/research/collection_online/

collection_object_details.aspx?objectId=1479568&partId=1&searchText=seal+89862&page=1.

———. 2022. "Cylinder Seal." *The British Museum.* Trustees of the British Museum. https://www.britishmuseum.org/collection/object/W_1888-0512-Bu-780.

Brockman, R. John. 1998. *From Millwrights to Shipwrights to the Twenty-First Century.* Cresskill, NJ: Hampton Press.

Cagirgan, G., and W. G. Lambert. 1991–1993. "Kislimu Ritual for Esagil." *Journal of Cuneiform Studies* 43/45: 89–106.

Canagarajah, Suresh. 2016. "Translingual Writing and Teacher Development in Composition." *College English* 78, no. 3: 265–73.

Carino, Peter J. 1996. "Open Admissions and the Construction of Writing Center History: A Tale of Three Models." *Writing Center Journal* 17, no. 1: 30–48.

Caswell, Nicole I., Jackie McKinney Grutsch, and Rebecca Jackson. 2016. *The Working Lives of New Writing Center Directors.* Logan: Utah State University Press.

Charbonnier, Georges. 1961. *Conversations with Claude Levi-Strauss.* London: Jonathan Cape.

Charpin, Dominique. 1995. *Ancient Near Eastern Art.* Berkley: University of California Press.

———. 2010a. *Reading and Writing Babylon.* Cambridge, MA: Harvard University Press.

———. 2010b. *Writing, Law, and Kinship in Old Babylonian Mesopotamia.* Chicago: University of Chicago Press.

Chicago Manual of Style. 2020. "5: Grammar and Usage 5.122: Imperative Mood." https://www.chicagomanualofstyle.org/home.html.

Cohen, Yoram, and Sivan Kedar. 2011. "Teacher-Student Relationships: Two Case Studies." In *Cuneiform Culture*, edited by Karen Radner and Eleanor Robson. Oxford: Oxford University Press.

Collon, Dominique. 1995. *Ancient Near Eastern Art.* Berkeley: University of California Press.

———. 1987. *First Impressions: Cylinder Seals in the Ancient Near East.* Chicago: University of Chicago Press.

Connor, H. T. H., and Jennifer J. Connor. 1992. "Commentary on Rhetorical Analysis of William Harvey's De Motu Cordis (1682)." *Technical Writing and Communication* 22, no. 2: 195–201.

Connor, Jennifer J. 1993. "Medical Text and Historical Context: Research Issues and Methods in History and Technical Communication." *Technical Writing and Communication* 23, no. 3: 211–32.

Cristomos, C. Jay. 2015. "Writing Sumerian, Creating Texts: Reflections on Text-building Practices in Old Babylonian Schools." *Journal of Ancient Near Eastern Religions* 15: 121–42.

Cultural Rhetorics Theory Lab. 2014. Michigan State University. Accessed January 2020. https://crtheorylab.wordpress.com/.

The Cursing of Agade: Translation. 2001. *ETCSLtranslation : t.1.1.2.* The Electronic Text Corpus of Sumerian Literature. https://etcsl.orinst.ox.ac.uk/section2/tr215.htm.

Delnero, Paul. Interview by Kathryn Raign. 2019. email, April 3.

———. 2010. "Sumerian Extract Tablets and Scribal Education." *Journal of Cuneiform Studies* 62:53–69.

Denny, Harry, Robert Mundy, Lilianna M. Haydan, Richard Severe, and Anna Sicari. 2018. *Out in the Center: Public Controversies and Private Struggles.* Edited by Harry Denny, Robert Mundy, Liliana M Naydan, Richard Severe, and Anna Sicari. Logan: Utah State University Press.

Dianakoff, J. M. 1986. "Women in Old Babylonia Not Under Patriarchal Authority." *Journal of the Economic and Social History of the Orient* 29, no. 2: 225–38.

Donahue, C. 2009. "Internationalization and Composition Studies: Reorienting the Discourse." *College Composition and Communication* 61, no. 2: 21–243.

Durack, Katherine. 1997. "Gender, Technology, and the History of Technical Communication." *Technical Communication Quarterly* 6, no. 3: 249–60.

Eisner, Will. 1996. *Graphic Storytelling.* Tamarac: Poorhouse.

Enos, Richard Leo. 1993. *Greek Rhetoric Before Aristotle.* Prospect Heights, IL: Waveland.

Escobar, Eduardo. Forthcoming. *Technology as Knowledge: Cuneiform Technical Recipes and the Material World.* Berkeley: University of California Press.

"ETCSLtransliteration : c.5.3.2." 2016. *The Electronic Web Corpus of Sumerian Literature.* Faculty of Oriental Studies. https://etcsl.orinst.ox.ac.uk/#.

Faculty of Oriental Studies. 2016. "Enki and Ninmah." *ETCSLtranslation : t.1.1.2.* The Electronic Text Corpus of Sumerian Literature. 11 1. https://etcsl.orinst.ox.ac.uk/cgi-bin/etcsl.cgi?text=t.1.1.2&charenc=j#.

———. 2021. "ETCSL Translation : t.5.1.3 The Advice of a Supervisor to a Younger Scribe (E-du-ba-a C)." *The Electronic Corpus of Sumerian Literature.* July 14. https://etcsl.orinst.ox.ac.uk/cgi-bin/etcsl.cgi?text=t.5.1.3#.

Falkowitz, Robert Seth. 1980. *The Sumerian Rhetoric Collection.* Ann Arbor, MI: University Microfilms International.

Finkel, Irving, and Jonathan Taylor. 2015. *Cuneiform.* Los Angeles: Getty.

Flynn, Elizabeth. 1997. "Emergent Feminist Communication." *Technical Communication* 6, no. 3: 313–18.

Gadotti, Alhena. 2014. "Sumerian Wisdom Literature." In *Women in the Ancient Near East*, edited by Mark W. Chavalas, 59–74. New York: Routledge.

Geller, Anne Ellen, Michelle Eddice, Frankie Condon, Meg Carroll, and Elizabeth H. Boquet. 2007. *The Everyday Writing Center.* Logan: Utah State University Press.

George, A. R. 2005. "In Search of the E.dub.ba.a: The Ancient Mesopotamian School in Literature and Reality." In *An Inexperienced Scribe Who Neglects Nothing: Ancient Near Eastern Studies in Honor of Jacob Klein* by P. A. Y. Sefati, 1–9. Bethesda, MD: CDL Press.

Goff, Beatrice L. 1956. "Amulets in Mesopotamian Ritual Rites." *Journal of the Warburg and Courtauld Institutes* 19 (1/2): 1–39.

Gorgias. 1990. "Encomium to Helen." In *The Rhetorical Tradition: Readings from Classical Times to the Present*, edited by Patricia Bizzell and Bruce Herzberg, 40–42. Boston: Bedford/St. Martin's Press.

Hallo, William J. 2004. "The Birth of Rhetoric." In *Rhetoric Before and Beyond the Greeks*, edited by Carol S. Lipson and Robert A. Brinkley, 1–261. Albany: State University of New York Press.

Halton, Charles, and Saana Svard. 2018. *Women's Writing of Ancient Mesopotamia: An Anthology of the Earliest Female Authors*. Cambridge: Cambridge University Press.

Harris, Rivkah. 1962. "Biographical Notes on the Naditu Women of Sippar." *Journal of Cuneiform Studies* 16, no. 1: 1–12.

———. 1964. "The Naditu Woman." In *Studies Presented to Leo A. Oppenheim*, 106–35. Chicago: University of Chicago Press.

———. 1975. *A Demographic Study of an Old Babylonian City (1894–1595)*. Vol. 36. Uitgaven, the Netherlands: Institut voor het Nabije Oosten.

Havelock, Eric A. 1963. *Preface to Plato*. Cambridge, MA: Harvard University Press.

"Henry the VIII: Defender of the Faith?" *Society of Antiquaries of London*. Society of Antiquaries of London. Accessed August 2022. https://stories.sal.org.uk/henryviii/object/mss-1008-a7/.

Hodges, Amy, Lynne Ronesi, and Amy Zenger. 2019. "Learning from/in Middle East and North Africa Writing Centers: Negotiaitng Access and Diversity." *The Writing Center Journal* 37, no. 2: 43–58.

Hunger, Hermann. 1968. "Babylonische und Assyrische Kolophone." In *Alter Orient und Altes Testament*, edited by Kevalaer, Butzon, and Bercker, 11. Neukirchen-Vloyn, Germany: Neukirchener Verlag des Erziehungsvereins.

Jacobsen, Thorkild. 2003. *The Context of Scripture: Canonical Compositions from the Biblical World*. Edited by William W. Hallo and K. Lawson Younger Junior. 3 vols. Boston: Brill.

Jarrat, Susan C. 1991. *Rereading the Sophists: Classical Rhetoric Refigured*. Carbondale: Southern Illinois University Press.

Jimenez, Enrique. 2020. "The Place of Disputation Poems within Babylonian Literature." In *Babylonian Disputation Poems*, 69–108. Berlin: De Gruyter.

J. P. Morgan Library. n.d. *Seal 747: Winged Hero Contesting Bull with a Lion for a Bull*. J. P. Morgan Library. http://www.themorgan.org/collection/ancient-near-eastern-seals-and-tablets/84369.

Kemp, Simon. 2022. *We Are Social.* December 1. https://wearesocial.com/au/blog/2022/04/more-than-5-billion-people-now-use-the-internet/.
Kennedy, George A. 1998. *Comparative Rhetoric: An Historical and Cross-Cultural Introduction.* New York: Oxford University Press.
Kiedaisch, Jean, and Sue Dinitz. 2007. "Changing Notions of Difference in the Writing Center: The Possibility of Universal Design." *Writing Center Journal* 27, no. 2: 39–59.
Kramer, Samuel Noah. 1949. "Schooldays: A Sumerian Composition Relating to the Education of a Scribe." *Journal of the American Oriental Society* 69, no. 4: 199–215.
———. 1981. *History Begins at Sumer: Thirty-Nine Firsts in Recorded History.* Philadelphia: University of Pennsylvania Press.
Kriwaczek, Paul. 2012. *Babylon, Mesopotamia and the Birth of Civilization.* New York: St. Martin's.
Lannon, John M., and Laura J. Gurak. 2017. *Technical Communication.* Boston: Pearson.
Lerner, Neal. 2009. *The Idea of a Writing Laboratory.* Carbondale: Southern Illinois Press.
Lu, Min-Zhan, and Bruce Horner. 2016. "Introduction: Translingual Work." *College English* 78, no. 3: 207–18.
Mao, LuMing. 2006. *Reading Chinese Fortune Cookie: The Making of Chinese American Rhetoric.* Logan: Utah State University Press.
McKinney, Jackie Grutsch. 2013. *Peripheral Visions of Writing Center.* Logan: Utah State University Press.
Meier, Samuel. 1991. "Women and Communication in the Ancient Near East." *Journal of the American Oriental Society* 111, no. 3: 540–47.
Melville, Sarah C. 2004. "Neo-assyrian Royal Women and Male Identity: Status as a Social Tool." *Journal of the American Oriental Society* 124, no. 1: 37–57.
Merriam-Webster Dictionary, s.v. "Eponymate," accessed November 1, https://www.merriam-webster.com/dictionary/eponymate.
Metropolitan Museum of Art. n.d. http://www.metmuseum.org.
Metropolitan Museum of Art. 2019. "Cylinder Seal: Seated Figure Approached by a Goddess Leading a Worshiper,ca. 2028–2004 B.C." *The Met*, August 19. https://www.metmuseum.org/art/collection/search/329060.
Monty, Randall W. 2016. *The Writing Center as Cultural and Interdisciplinary Contact Zone.* Palgrave Macmillan.
Neel, Jasper. 1988. *Plato, Derrida and Writing.* Carbondale: Southern Illinois University Press.
Nissinen, Martti. 2013. "Gender and Prophetic Agency in the Ancient Near East and in Greece." In *Prophets Male and Female: Gender and Prophecy in the Hebrew Bible, the Eastern Mediterranean and the Ancient Near East*, edited

by Jonathan Stokl and Corrine L. Carvalho, 27–58. Atlanta: Society of Biblical Literature.

Norman, Don. 2013. *The Design of Everyday Things.* New York: Basic Books.

Oinonen, Virpi. 2016. "Why Cartoons are Powerful Communication Tools." *Business Illustrator.* November 12. http://www.businessillustrator.com.

Ong, Walter. 1975. "The Writer's Audience is Always a Fiction." *Publications of the Modern Language Association* 90, no. 1: 9–21.

Oppenheim, A. Leo. 1964. *Ancient Mesopotamia: Portrait of a Dead Civilization.* Chicago: University of Chicago Press.

———. 1967. *Letters from Mesopotamia.* Chicago: University of Chicago Press.

Pardee, Dennis. 2003. *Ilu on a Toot.* Vol. 1 of *The Context of Scripture*, edited by William W. Hallo and K. Lawson Younger, 302–5. Boston: Brill.

Parker, Barbra. 1961. "Administrative Tablets from the North-west Palace, Nimrud." *Iraq* 22, no. 1: 15–67.

Pearce, Laurie E. 1995. *The Scribes and Scholars of Ancient Mesopotamia.* Vol. 4 of *Civilizations of the Ancient Near East.* Edited by Jack M. Sasson. New York: Charles Scribner's Sons.

Plato. 1956. *Phaedrus.* Edited by W. C. Hembold and W. G. Rabinowitz. Translated by W. C. Hembold and W. G. Rabinowitz. Upper Saddle River, NJ: Prentice Hall.

———. n.d. *Sophist.* Edited by Benjamin Jowett. Kindle.

Ponchia, Simonetta. 2007. "Debates and Rhetoric in Sumer." In *Traditions of Controversy*, edited by M. Daschal and H. Chang, 63–83. Philadelphia: John Benjamin.

Porada, Edith. 1980. "Introduction." In *Ancient Art in Seals*, edited by Pierre Amiet, Nimet Ozguc, and John Boardman, 3–20. Princeton, NJ: Princeton University Press.

Powell, Barry M. 2012. *Writing: Theory and History of the Technology of Civilization.* Oxford: Wiley-Blackwell.

Powell, Malea. 2002. "Rhetorics of Survivance. How American Indians Use Writing." *College Composition and Communication* 53, no. 3: 396–434.

Powell, Malea, Daisy Levy, Andrea Riley-Mukavetz, Marilee Brooks-Gillies, Marie Novotny, and Jennifer Fisch-Ferguson. 2014. *Our Story Begins Here: Constellating Cultural Rhetorics.* October 25. Accessed January 24, 2020. http://enculturation.net/our-story-begins-here.

Protagoras. 1972. "Concerning the Gods." In *The Older Sophists*, edited by Rosamund Kent Sprague and translated by Michael J. O'Brien. Columbia: University of South Carolina Press.

Raign, Kathryn Rosser. 2013. "Creating Verbal Immediacy: The Use of Immediacy and Avoidance Techniques in Online Tutorials." *Praxis: A Writing Center Journal* 1, no. 2.

———. 2019. "Finding Our Missing Pieces: Women Technical Writers in Ancient Mesopotamia." *Journal of Technical Writing and Communication* 49, no. 3: 338–64.
Ratcliffe, Krista. 2010. "The Twentieth and Twenty-First Centuries." In *The Present State of Scholarship in the History of Rhetoric: A Twenty-first Century Guide*, edited by Lynee Lewis Gaillet and Winifred Bryan Horner, 185–212. Columbia: University of Missouri Press.
Robson, Eleanor. 2001. "The Tablet House: A Scribal School in Old Babylonian Nippur." *Revue D'Assyriologie et D'Archeologie Orientale* 93, no. 1: 39–66.
Royster, Jacqueline Jones, and Gesa E. Kirsch. 2012. *Feminist Rhetorical Practices: New Horizons for Rhetoric, Composition, and Literacy Studies*. Carbondale: Southern Illinois University Press.
Schmandt-Besserat, Denise. 2016. "3000–2600 BC: How Writing Came to Replicate Spoken Speech." Rutgers University, January 27.
———. 1996. *How Writing Came About*. Austin: University of Texas Press.
———. 2007. *When Writing Met Art: From Symbol to Story*. Austin: University of Texas Press.
Schwemer, Daniel. 2018. "The Ancient Near East: Magic: Origin and Meaning." In *The Cambridge History of Magic and Witchcraft in the West from Antiquity to the Present*, edited by David J. Collins, 17–51. Cambridge: Cambridge University Press.
Scurlock, JoAnn. 2014. *Sourcebook for Ancient Mesopotamian Medicine*. Vol. 36. Atlanta: Society of Biblical Literature.
Sjoberg, A. W. 1975. *Sumerological Studies in Honor of Thorkild Facobsen*. Edited by S. J. Lieberman. Chicago: University of Chicago Press.
Snow, C. P. 1961. *The Two Cultures and the Scientific Revolution*. New York: Cambridge University Press.
Stol, Marten. 2016. *Women in the Ancient Near East*. Berlin: De Gruyter.
Stone, Elizabeth. 1982. "The Social Role of the Naditu Women in Old Babylonian Nippur." *Journal of the Economic History of the Orient* 25, no. 1: 50–70.
Swearingen, Jan. 1986. "Literate Rhetors and Their Illiterate Audiences: The Orality of Early Literacy." *PRE/TEXT* 3: 145–62.
Taylor, Irving, and Jonathan Finkel. 2015. *Cuneiform*. Los Angeles: Getty.
Tebeaux, Elizabeth. 1998. "The Voices of English Women Technical Writers, 1641–1700: Imprints in the Evolution of Modern English Prose Style." *Technical Communication Quarterly* 7, no. 2: 125–52.
———. 2009. "The Association of Teachers of Technical Writing: The Emergence of Professional Identity." *Technical Communication Quarterly* 18, no. 2: 107–41.
Teissier, Beatrice. 1984. *Ancient Cylinder Seals from the Marcopoli Collection*. Berkeley: University of California Press.

Tetlow, Elisabeth. 2004. *Women, Crime and Punishment in Ancient Law and Society.* New York: Continuum.

Tudeau, Johanna. 2016. "Nidaba (Goddess)." In *Ancient Mesopotamian Gods and Goddesses.* Edited by Oracc and UK Higher Education. London: UK Higher Education. http://oracc.museum.upenn.edu/amgg.listofdeities/nidaba.

Vanstiphout, H. L. J. 1990. "The Mesopotamian Debate Poems. A General Presentation. Part I." *Acta Sumerologica* 12: 271–318.

———. 1991a. "Lore, Learning and Levity in the Sumerian Disputations." In *Dispute Poems and Dialogues in the Ancient and Mediaeval Near East,* by G. J. Reinink and H. L. J. Vanstiphout, 23–46. Leuven: Peeters.

———. 1991b. *Dispute Poems and Dialogues in the Ancient and Mediaeval Near East: Forms and Types of Literary Debates in Semitic and Related Literatures.* Edited by G. J. Reinink and Herman L. J. Vanstiphout. Vol. 42. Leuven: Orientalia Lovaniensia Analecta.

———. 1992. "The Mesopotamian Debate Poems. A General Presentation. Part II: The Subject." *Acta Sumerologica* 14: 339–68.

———. 1997. "School Dialogues in the Context of Scripture." In *Canonical Compositions from the Biblical World,* edited by W. W. Hallo and K. I. Younger, 588–93. New York: Brill.

———. 2003a. *Enmerkar and the Lord of Aratta.* Edited by Jerrold S. Cooper. Vol. 20. Atlanta: Society of Biblical Literature. https://etcsl.orinst.ox.ac.uk/section1/tr1823.htm.

———. 2003b. *The Disputation Between Bird and Fish.* Vol. 1 of *The Context of Scripture.* Edited by William W. Hallo. Boston: Brill.

———. 2003c. "The Disputation Between the Hoe and the Plow." In *The Context of Scripture: Canonical Compositions from the Biblical World,* edited by William W. Hallo and K. Lawson Younger Jr., 578–81. Boston: Brill.

Walton, Rebecca, Kristen R. Moore, and Natasha N. Jones. 2019. *Technical Communication After the Social Justice Turn: Building Coalitions for Action.* New York: Routledge.

Wikipedia. 2019. "Sumer, Akkad, and Elam." Accessed April 23, 2023. https://www.google.com/search?as_st=y&tbm=isch&as_q=map+of+ancient+mesopotamia&as_epq=&as_oq=&as_eq=&cr=&as_sitesearch=&safe=images&tbs=ift:jpg,sur:f#imgrc=XQhrEnaQ6WSo7M.

Wiseman, D. J. 1977. *Cylinder Seals of Western Asia.* London: Batchworth.

Wolkenstein, Diane, and Samuel Noah Kramer. 1983. *Inanna: Queen of Heaven and Earth.* New York: Harper & Row.

Young, Iris. 2003. "Five Faces of Oppression." In *Oppression, Privilege, and Resistance,* by Lisa Heldke and Peg O'Connor, 37–63. Boston: McGraw-Hill.

Yu, Han. 2017. *The Other Kind of Funnies: Comics in Technical Communication.* Edited by Charles Sides. New York: Routledge.

Index

accomplish, 67–69, 71–72, 83–84
accounting, 8, 10, 59, 79, 85, 93, 104
adjective, 52
adjectives, 52, 125
Adjudication, 116
adjudication, 114, 116, 123
administrative, 10–11, 38–39, 47–48, 51, 64, 68–69, 96, 99, 108
adversative, 49, 51
Agade, 79, 165
agency, 84
agent, 68–69
agrarian, 79, 131
Akkadian, 1–2, 9–10, 28, 33, 84, 94
Alster, 64, 74, 103, 105, 163
Amiet, 168
Anatolia, 146
anatomical, 136
antinarrative, 14
archetypal, 11
Argon, 29
argue, 19, 29–30, 41, 73, 91–92, 107, 115, 119
argument, 44, 48, 63, 71, 81–82, 92, 114–117, 119, 121–123, 127–128, 130–131, 161
argumentation, 161
Aristotle, 58, 70, 81, 91, 114, 120, 165
art, vii, 3, 19–20, 38, 80, 102, 104, 119–120, 151, 167

asipu, 134–135, 146
Assur, 31, 42, 45–46, 48–49, 51, 82–83
Assurbanipal, 82–83
Assyria, 10, 31, 65, 82–83, 163
Assyrian, 1, 9–10, 30–31, 149
Assyriologie, 169
asu, 76, 134–135, 145–146

Babylon, 1, 85, 150, 163–164, 167
Babylonia, 65, 67, 163, 165
Babylonian, 9, 30–31, 47–48, 66, 74–76, 78, 84–85, 107–109, 111, 120, 145, 147, 164, 166, 169
banquet, 27, 86
Bayliss, 85, 163
Bazerman, 161
BC, 9–10, 156, 169
Bellos, 7, 163
Besnier, 96, 163
Besserat, 4, 9, 26–27, 85–86, 169
Bird, 11, 115, 117, 120–124, 170
Bizzell, 117, 163, 166
Boje, 14, 163
Boquet, 97, 109, 163, 165
Bratta, 6–7, 163
Brauer, 103, 112, 161–163
Brockman, 39, 59, 164
Brooks, 4, 168
Bryan, 169
bulla, xi, 19

172 | Index

bullae, 19
business, 2, 9–11, 20–21, 24, 35, 61–62, 65–70, 78, 80, 84–85, 87, 96, 157

Canagarajah, 94, 164
canon, 2, 74, 76, 135, 153
Carino, 109, 164
Carlino, 162
Carroll, xvii, 165
Carvalho, 168
causal, 127
ceremony, 85
chant, 3, 95
Charbonnier, 150, 164
Charpin, 28, 33–34, 68, 79, 84, 97, 99, 108–109, 147–149, 155, 158–159, 164
Chavalas, 165
chief, 41, 102, 152
childbirth, 149
Chinese, 5–6, 62, 167
Cicero, 70, 120
citizen, 95, 97, 155–156
Clarendon, 162
clause, 40, 43–46, 49, 51, 54, 161
cloister, 64, 77
Cohen, 100–101, 164
collaborate, 48, 99, 159
collaboration, 107
Collins, 169
Collon, 17, 23, 26, 28, 76, 164
colophon, 41, 45, 144
combatants, 20, 30
commodity, 9, 19, 21
communicate, 16, 18, 20–21, 39, 87, 113
communication, xv–xvi, 1, 3, 6, 11–12, 17–18, 22, 27, 35, 38, 59, 62, 64, 78, 86–87, 105, 108, 147–148, 150, 157

community, 61, 64, 75, 97, 105, 120, 123, 134, 147, 150
complement, 141
complex, 8, 13, 28, 30, 48, 61, 67–68, 71, 75, 89, 91, 97, 106, 118, 131, 145, 150
composition, xv–xvi, 79, 83, 90, 93, 109, 151
conditional, 38, 40, 43–45, 49, 54–57, 108
conjunction, 43, 86
Connor, 7–8, 164, 170
consumer, 148
contrastive, 51
counterarguments, 116
counterpoint, 5
Courtauld, 166
crawl, 71
creditor, 84–85, 159
Cresskill, 164
Cristomos, 107, 164
cultivation, 79
culture, 6–8, 12, 17, 19, 25, 29, 34–35, 38, 44, 62–64, 66–69, 73, 79, 89, 97, 105, 120–121, 150–151, 154
Cuneiform, xii, 1–3, 6–9, 11, 13, 15, 17, 38, 61, 63, 68, 89, 91, 93, 95–96, 106, 110–111, 118, 132, 138, 150, 158, 163–167, 169
curriculum, 95
cylinder, vii, xii, xvii, 10, 13–14, 16, 18–24, 28, 30–31, 34–35, 76, 78, 84, 111, 150, 154, 157
cylinders, 23

dead, 4, 6, 11, 27, 61, 81, 85–86, 158
death, 31–32, 75, 160
debate, 114–116, 120
debates, 97, 124
debt, 50, 69, 159
debtors, 49–50, 84, 159

debts, 159
deity, xi, 31, 71, 79–80, 102
Delnero, 94–95, 99, 107, 109, 111, 165
demon, xvi, 72, 136, 152
Denny, 92, 165
dependent, 46, 65, 84, 125, 128, 131
Derrida, 167
design, xvi, 13, 17–18, 20, 22, 134
dialect, 9
dialogue, 114–115, 123
dialogues, 11, 107, 114, 116
Dianakoff, 66–67, 165
Dinitz, 92, 167
discourse, 14, 92, 119–120, 161
disk, 31
disks, 30
disputation, 11, 114, 117, 121, 123, 125, 127, 131–132, 153
disputations, 3, 11, 113–124, 132
disputatious, 119
disrupts, 45, 55
Dissoi, 162
dissoi, 123
ditto, 140, 142
doctor, 35, 76, 134, 137, 141, 143, 145–146
doctors, 70, 76–77, 134–137, 143–146
document, 15, 17, 22, 39, 66–67, 70, 76, 78, 85, 94, 97, 101, 103, 109, 134, 143
documented, 5
documenters, 148
documents, xvi, 2–3, 6–11, 37–39, 44, 59, 62–65, 69–70, 80, 87, 97, 101, 107, 132, 139, 143, 159
Donahue, 91, 161–162, 165
dowry, 65–67, 161–162
drilled, 18
Durack, 61–62, 68–70, 72–73, 165
dynastic, 20

Eduba, 107
Edubba, ix, xiii, 81, 89, 93, 109
edubba, 11, 64, 90–91, 94, 97, 99, 101, 103, 105–109
edubbas, 99
education, 6, 11, 63, 65, 82, 97–99, 101, 106–107, 111, 118, 148, 151
egalitarian, 154
Egypt, 94, 109
Egyptian, 62, 79
Eisner, 165
elite, 79, 127, 153
emotions, 20, 83, 125
Enheduana, 9, 11, 107
Enheduanna, 63–64, 69–73, 105
Enki, 122, 124, 151–153, 165
Enlil, 72, 81, 122, 124, 129, 131
Enmerkar, 121, 170
Enos, 11, 38, 119, 161, 165
eristic, 119–120
Escobar, 41, 165
ethical, 41, 117, 125
ethos, 41, 114–115, 117, 121, 124–125
evidence, 11, 26, 39–40, 53, 59, 62–65, 67–68, 70, 73, 76, 78–79, 81, 105–109, 117, 145, 149, 154, 159
Ewe, 118, 122–124

Falkowitz, 120, 165
female, 30, 34, 62–63, 65, 68, 70, 73, 76–77, 82, 145–146, 160
Ferguson, 4, 168
figure, 17, 28–30, 33, 70, 86, 90, 94
Finkel, 95, 165, 169
Fish, 11, 115, 117, 120–124, 163, 170
flattening, 94
Flynn, 86–87, 165
Freiberg, 112

Gadotti, 74, 165
Geller, 6, 89, 103, 165

gender, 11, 62–63, 67–68, 148
George, 64, 91, 99, 104–105, 165, 167
Gerd, 112, 161, 163
Gesa, 92, 169
Gilgamesh, 2, 9, 15, 29–30, 75, 95, 97, 155
Giroux, 163
Goetze, 53
Goff, 12, 21, 76–78, 166
Gorgias, 115–116, 166
grant, 3
Greece, 11, 38, 80, 92, 119, 167
Greek, 11, 114, 120, 134, 161, 165
Greeks, xvi, xviii–3, 39, 59, 71, 81, 87, 90–91, 114–115, 131–132, 166
Grutsch, 92, 99–100, 164, 167
Gruyter, 166, 169
Gurak, 17, 167

hairs, 133
Hallo, 119, 166, 168, 170
Halton, v, 1, 10, 38, 40–41, 44, 61–63, 68–70, 72, 76, 82–83, 108, 166
Hammurabi, 30, 64, 84, 105, 134, 145, 149, 159
Hampton, 164
handwriting, 18, 22
hangover, 133
Harper, 170
Harris, 64, 105–107, 166
Havelock, 116, 127, 166
healers, 146
Hebrew, 167
Herennium, 114
Herodotus, 134, 145
Herzberg, 117, 163, 166
Hetaerae, 67
hetaerae, 66
heuristic, 119
Hittite, 94
Hoe, xii, 11, 95, 113, 117–118, 121–123, 125–131, 153, 162, 170
Homer, 3, 11, 38, 63, 70, 76, 81

Homeric, 116, 119, 122, 126, 162
Horner, 93, 161–162, 167, 169

iconographic, 28
iconography, 19–20, 29, 33
ideological, 15, 17, 21–22, 28, 33, 85, 150
Ideology, 21
imperative, 17, 38–40, 44–45, 47–49, 51–52, 55, 59, 77, 108, 130
Inana, 95
Inanna, 11, 25–26, 63–64, 67, 69, 71–72, 79, 105, 170
Incantation, 77–78
incantation, 78
independent, 43, 46, 61, 69–70, 150
indicative, 40, 44, 79
indirect, 47, 51
infinitive, 51
instructional, 3, 10, 38–39, 58–59, 62–63, 71, 87, 94
instructions, xvi, 1, 37–39, 41, 43, 45–52, 55, 58–59, 64, 69–73, 76–80, 87, 98, 103, 108, 116, 120, 127, 136, 146, 160
instructor, 1, 94–95, 98, 109, 111
instructors, 98–99, 105, 112
Iran, 13
Iraq, 13, 163, 168

Jacobsen, 166
Jarratt, 120, 126
Jasper, 117, 167
Jimenez, 118, 122–123, 166
Jowett, 168

Kalki, xi, 14, 28–29
Kearns, 162
Kedar, 100–101, 164
Keith, 162
Kemp, 148, 166
Kennedy, 91, 167
Kent, 168

Kevalaer, 166
Kiedaisch, 92, 167
Kimball, ii
kinfolk, 103
King, 2, 29, 33, 70, 81–82, 128, 135, 163
king, 4, 13, 15, 17, 26, 29, 31, 33–34, 41, 45, 64–66, 74, 81–84, 96–97, 105–106, 117, 122, 125, 128, 130, 149, 152–153, 155–156
kings, 13, 17, 34–35, 68, 96, 128, 130, 147–149, 154–155
Kirsch, 5–6, 91–92, 169
Klein, 166
Kramer, 11, 25, 80, 101, 103, 156, 167, 170
Kriwaczek, 1–2, 134, 145, 150–151, 154–155

Lambert, 74, 78, 167
Lannon, 17, 167
Lawson, 166, 168, 170
lemmata, 107
lentil, 89, 94
Lerner, 90–91, 167
letter, 2–3, 10–11, 37–38, 47–49, 51–54, 57–58, 64, 69, 81–84, 105, 108–109, 158–159
letters, xvi, 2–3, 8–10, 38–39, 47–48, 53, 59, 69–70, 80, 82–84, 95, 106–108, 120
Leuven, 170
Lewis, xvii, 169
Lieberman, 169
Lipson, 166
literacy, 6, 35, 63, 68, 149, 157
literate, 68, 120, 149, 153, 158
literature, xvi, 2, 8–9, 11, 25, 62, 64, 68–69, 74, 99, 105–107, 116
logical, 10, 27, 37, 55, 82–83, 86, 118, 122–123, 130, 157
logographic, 93
logoi, 123

logos, 11, 114, 117–118, 121–122, 127, 132
Lugal, 155
Lundsford, 161

MacAloon, 163
MacKay, 162
manual, 10, 37–39, 41, 45, 47
manuals, 8, 10, 38–40, 44, 59, 108, 160
manuscripts, 118
Mao, 5–7, 162, 167
map, 112, 170
Marcopoli, 169
marriage, 61, 151
Martin, 84, 166–167
Martins, 162
mashmashu, 12
Matsuda, 162
medical, 38, 76, 134–135, 137–138, 141–145, 160
medication, 134, 146
medicine, 12, 59, 96, 134, 145–146
Meier, 63–64, 105, 167
Melville, 65, 167
memory, 14–15, 104, 107
mentor, 98
merchant, 133
Mesoamerican, 62
Mesopotamia, v, vii, ix–x, xiii, xvii, 1, 3, 5, 8, 10–12, 20, 34, 37, 44, 47, 59, 61, 63, 67, 69, 80, 84, 87, 121, 131, 133, 138, 146–148, 150, 157, 164, 167–169
Mesopotamian, ix, xii–xiii, xvi, 2, 9, 25, 39, 62–63, 67–70, 73–74, 79, 89–90, 98, 114–116, 134–136, 145, 151, 165–166, 169–170
methodology, 7, 92
military, 68, 82, 149, 155
monolingualism, 93
mood, 38, 40, 44–45, 47
Moore, 148, 157, 170

Mukavetz, 4, 168
Myers, 162
myth, 12, 121, 134
mythic, 15, 121
mythos, 11, 114, 117–118, 121, 132
myths, 15, 95, 97, 121–122

naditu, 64–65, 67, 74, 77–79, 105–107, 161–162
Naditum, 66
naditus, 65, 106
Neal, 90, 167
Neel, 117, 167
Nidaba, 64, 79–80, 102–103, 105, 170
Ninsatapada, 81–82
Ninurta, 67
Nippur, 66–67, 99, 118, 169
Nisaba, 98, 116
Nishatapada, 64, 105
Nnmrud, 168
nomoi, 15, 116, 127, 131
nomos, 11, 114, 117–118, 120–121, 126, 132
Novotny, 4, 168

Ong, 116, 168
Oppenheim, 2, 4, 6, 12, 38, 40, 47–48, 53, 58, 82, 93–94, 108, 149, 158, 161, 166, 168
orality, 6, 149

passive, 45, 47, 52, 58
pathos, 114, 117, 121, 124
patriarchal, 65–67, 73–74, 91
Pearce, 64, 96, 105, 168
Pearson, 167
pedagogy, 109
Peeters, 170
Perfume, 3, 41, 43, 45
persuasion, xvi, 3, 39, 52, 59, 71, 113–114
Phaedrus, 168

pharmacist, 146
pharmacology, 139
pharmacy, 146
phonographic, 93
phrase, 19, 58, 65, 106–107, 128
pictographic, 10
plants, 38, 134, 136, 139, 141–145
Plato, 70, 79, 91, 114, 116, 119–120, 166–168
Plow, 11, 113, 117–118, 121–123, 125–131, 153, 162, 170
Ponchia, 117–119, 121, 124, 168
Porada, 23, 168
possession, 157
Powell, 4, 6–7, 10, 37, 91, 163, 168
praxis, 92
preliterate, 122
prepositions, 40
priest, 12
priestess, 64, 66, 73, 77–79, 81
procedure, 116
procedures, 135, 160
propaganda, 28
prose, 41, 64, 105
prostitute, 66–67
Protagoras, 117, 168
protreptic, 119–120
proverb, 96
Puabi, xi, 26–27, 33, 70, 85–86, 155

queen, 27, 33, 64, 66, 71, 80, 84, 86, 102–103, 105, 130, 152–153, 155
Quintilian, 70, 120

Rabinowitz, 168
Radner, 164
Raign, xiii, 98, 165, 168
Ratcliffe, 92, 169
read, xvii, 1, 7, 18, 39, 43, 61, 68, 73, 86, 90, 101, 149, 153, 158
reader, 17, 31–32, 39, 43–45, 50, 53, 58, 69, 86, 116, 133, 143, 160

reading, xvii, 2–4, 39, 44, 93, 158
receipt, 2, 13, 147
receipts, xvi, 9–10, 70, 80, 84
recipe, 41, 43, 45, 47, 134
recipes, 3, 37, 44, 108
recordkeeping, 150
redress, 84
reed, 72, 77, 80, 89, 102, 126, 129, 138
refutation, 114
refute, 115
refuting, 115
Reinink, 170
repetition, 50, 84, 118, 127, 138, 141–142
rhetoric, xv–xvi, 4–6, 11, 38–39, 53, 59, 91, 114, 117, 119–121, 131–132
rhetorical, xv–xvi, 3, 5–6, 11, 39, 45, 47, 51, 57, 59, 63, 71, 81, 84, 91, 108, 114–121, 123, 132
rhetorically, 6, 39
rhetorician, 114
rhetoricians, 91
rhetorics, 5, 91–92
Rich, 119
Riley, 4, 168
Robson, 99, 164, 169
Rogers, 161
Royster, 5–6, 91–92, 169

salutation, 108
Sargon, 28, 33, 64, 70, 105
Sarrat, 82–83
Sasson, 168
Schmandt, 4, 9, 26–27, 85–86, 169
school, 5–6, 33, 59, 64, 79, 81, 90, 98–106, 113
Schwemer, 146, 169
Scott, 161–162
scribal, 3, 6, 8, 10, 28, 65, 68, 73, 80–81, 83, 90, 95, 97, 101–104, 106–107, 111, 117–118, 120, 132

scribe, 4, 11, 13–14, 28, 34, 37, 41, 48, 63–65, 67–68, 81, 84, 96–99, 104–110, 138, 149, 158–159
scribes, 2–3, 9, 11, 28, 48, 63–65, 68, 70, 79–80, 84–85, 90, 93, 96–97, 99, 105–107, 118, 132, 139, 149, 158–159
Scurlock, 76, 135–137, 139–145, 169
seal, vii, xi, xvii, 10, 13–24, 26–35, 53–54, 56, 58, 67, 76, 78, 84, 86, 147, 157–158, 164
sealing, 19, 78, 84
seals, xvii, 10, 13–24, 26, 28–31, 33–35, 78, 84, 86, 150, 154, 157–158, 166
sekrum, 70, 74–75
sequential, 138, 143
series, ii, 61, 75, 95, 119, 127, 135–137, 139, 143, 155
signature, xi, xviii, 10, 13, 15, 18–22
signed, 13, 84
Sippar, 64–67, 106–107, 166
Sjoberg, 5, 96–97, 169
Socrates, 91
sophist, 117
sophistic, 11, 114, 116–117, 120–121
sophistry, 119
sophists, 3, 114, 117, 119–121, 123–124
Sprague, 168
steps, 40, 42, 44, 46, 55, 58–59, 78, 114
Stokl, 168
Stol, 26, 76, 84, 134–135, 145, 149, 161, 169
Stone, 22, 67, 169
Strauss, 164
student, 5, 11, 53, 59, 68, 92, 94–98, 100–105, 109–111, 135, 137
students, xv–xvi, 3, 59, 73, 90, 92–95, 97–101, 103, 105–109, 112, 118, 120, 134, 143, 160

style, xvi, 33, 39, 47, 63, 97, 115
styles, 24
stylistic, 19, 47–48, 128
stylus, 68, 95, 138
subjective, 57
subjugated, 51
Sumer, 2, 12, 27, 53, 65, 85–86, 119, 134, 163, 167–168, 170
Sumerian, 1, 7, 9–10, 25, 29, 47, 63–64, 71, 79, 84, 94, 97, 99, 101, 104–105, 107–108, 117, 119–120, 123, 133, 163–165, 167, 170
Sumerians, 17, 29, 73, 97, 121, 132
Sumerological, 169
suppression, 33–34
Svard, v, 1, 10, 38, 40–41, 44, 61–63, 68–70, 72, 76, 82–83, 108, 166
Swearingen, 120, 169
syllabaries, 94

tablet, v, xii, 1, 5, 12, 28, 41, 53–56, 64, 66–68, 83–86, 89, 91, 94–95, 101, 104, 110–111, 122, 133, 138, 141–143, 145, 158–159
Tablets, xii, 94, 110–111, 165, 168
tablets, xvi, 2, 6, 10–11, 26, 41, 59, 67, 84, 86, 89, 94–97, 101, 103, 107, 109, 112, 131, 135–136, 138, 141–143, 145, 149, 159, 166
Tapputi, 39, 41, 43, 45
teach, xv, 3, 29, 39, 59, 62, 98, 103, 112, 117–118, 120–121, 123, 143, 160
Teacher, 104, 164
teacher, 6, 61, 89, 95, 98–99, 101–104, 109–110, 120
teachers, 11, 59, 93, 98, 100–101, 105, 120
teaching, xv, 3, 10–11, 81, 90–91, 93, 100–102, 107, 109, 114, 121, 162
Tebeaux, 3, 62–63, 169
techne, 119, 131

technical, xv, xvii, 1, 3, 7–8, 10, 12, 14, 17–18, 20–21, 29, 34, 38–39, 41, 43–45, 47, 61–64, 67–71, 73, 79, 85–87, 96–97, 118, 147
technicality, 40
technically, 13
techno, 157
technological, 10, 118, 121–122, 148
technology, xviii, 12, 22, 59, 86–87, 131
Teissier, 18–20, 28–29, 169
tell, 5–6, 13, 24, 28, 35, 37, 92, 97, 101, 104, 107, 116, 143
Tetlow, 64–65, 81, 105, 170
texts, xvi–xvii, 3, 6–8, 10–12, 18, 26, 29, 39–41, 59, 62–64, 67, 69–70, 74–75, 79, 82, 85–87, 95, 103, 106–107, 109, 134–136, 138, 141, 143, 149
textual, 11, 39, 62–64, 67–68, 76, 79, 105–106, 145
theocratic, 151
therapeutic, 146
There, 19, 118, 141, 150
Theuth, 79
Thompson, 61–62
Thorkild, 166, 169
topoi, 119
Transactions, 62
transatlantic, 92
transcribe, 85, 158
translate, 48, 159
translingual, 93, 162
translingualism, 93
transliteration, 139
treatments, 76, 136–137, 143, 145
Trimbur, 162
Tudeau, 79, 170
tutor, 92–93, 104, 107
tutored, 109
tutorials, 103
tutoring, 92, 105, 109

tutors, 92–93, 98, 100, 103–105, 112

Uruk, xi, 10, 24, 75, 81, 85, 109, 154

Vanstiphout, 113, 115–124, 126–128, 131, 170

Walton, 148, 157, 162, 170
warlords, 154
Wheat, 122–124
Winter, 57, 122–123
Wiseman, 15, 30–32, 170
witches, 146
Wolkenstein, 25, 170
woman, 4, 11, 22, 24, 26, 33–34, 41, 61, 63–70, 72–77, 79, 82, 84, 105, 116, 145–146, 149, 152, 158–159
women, 11, 14, 23, 25, 33–35, 39, 61–74, 76–80, 82, 84–87, 100, 105–106, 108, 145, 148, 155, 161–162

wrangle, 113, 120
wrangles, 119–120
write, 4, 6, 10–11, 13, 37, 39, 53, 61–62, 65, 67–69, 73, 78–80, 82–84, 87, 89–90, 93, 96–97, 99, 104, 106–109, 112, 120, 135, 138, 143, 149, 158, 160
writings, 62, 132
written, iv, vii, 3–6, 9–11, 16, 21, 27, 29, 38, 41, 45, 47, 53, 58–59, 62–65, 69–70, 73, 75–76, 80, 83, 85–87, 95, 101, 106–109, 114–115, 123, 132–135, 143, 145, 147–149, 155, 157, 159–160

Young, 74, 80, 102, 149, 153, 170
Yu, 18, 170

Zainab, 62, 163
zakır, 3, 27, 85–86
Zhan, 162, 167

www.ingramcontent.com/pod-product-compliance
Lightning Source LLC
Chambersburg PA
CBHW022028240426
43667CB00042B/1408